Applications of Minimally Invasive Nanomedicine-Based Therapies in 3D in vitro Cancer Platforms

Synthesis Lectures on Materials and Optics

Applications of Minimally Invasive Nanomedicine-Based Therapies in 3D in vitro Cancer Platforms
Layla Mohammad-Hadi and Marym Mohammad-Hadi

ISBN: 978-3-031-01260-0   paperback
ISBN: 978-3-031-02380-0   ebook
ISBN: 978-3-031-00252-6   hardcover

DOI 10.1007/978-3-031-02380-0

A Publication in the Springer series
SYNTHESIS LECTURES ON MATERIALS AND OPTICS

Lecture #6
Series ISSN
Synthesis Lectures on Materials and Optics
Print 2691-1930   Electronic 2691-1949

Applications of Minimally Invasive Nanomedicine-Based Therapies in 3D in vitro Cancer Platforms

Layla Mohammad-Hadi and Marym Mohammad-Hadi
University College London

SYNTHESIS LECTURES ON MATERIALS AND OPTICS #6

&

ABSTRACT

Minimally invasive techniques such as Photodynamic Therapy (PDT) and Photochemical Internalisation (PCI) have for years been under investigation for the treatment of solid cancers. A significant number of the recent research studies have applied PDT and PCI to biological three-dimensional (3D) cancer platforms with many of the studies also involving the use of nanoparticles in order to enhance the efficacy of these light-based therapies. Interest in the employment of 3D cancer platforms has increased considerably due to the ability of the platforms to mimic *in vivo* models better than the conventional two-dimensional (2D) cultures. Some of the advantages of the 3D cancer systems over their 2D counterparts include improved interaction between cancer cells and the surrounding extracellular matrix (ECM) as well as restricted drug penetration which would allow optimization of treatments prior to undertaking of *in vivo* studies. The different chapters of this book will discuss photosensitizers and nanoparticles used in PDT and PCI in addition to the applications of these treatments in various 3D cancer models.

KEYWORDS

light therapy, non-invasive medicine, 3D models, nanomedicine, photodynamic therapy, photochemical internalization

We dedicate this book to our beloved parents
for their endless love, support, and sacrifice over the years.

Contents

Preface

In recent decades, light-based therapies have emerged as promising minimally invasive techniques for tackling various diseases. The combination of nanomedicine with such therapies has created many possibilities for enhancing the efficiency of these treatments and has thus been subject of immense research, which has resulted in encouraging developments.

Similarly, the introduction of different three-dimensional (3D) cancer constructs to research and their utilization as platforms in anti-cancer studies, including those involving light-based therapies has provided researchers with a model system which is reproducible and easy to create. Such models allow different densities and components to be used with the construct to match the purpose of the research better. Moreover, the 3D models provide an opportunity for the generation of complex cancer constructs that can be used in studies focusing on cancer behavior as well as various anti-cancer treatments including those which involve nanoparticles. The capability to develop complex cancer constructs using such platforms may furthermore imply that the 3D models have the potential to contribute toward reducing the use of animal models in cancer-related studies in the future which in turn could also reduce animal suffering. Additionally, the cancer models can be developed using tissue samples from the patient to simulate the clinical tumor so that the response of the cancer to different therapeutics can be tested quickly. This allows the use of personalized cancer treatment instead of subjecting the patient to different treatments, with significant benefits to the quality of the patient's life. The models also allow the incorporation of immunological factors which enables testing for any side effects or allergic reactions that may occur as a result of using the therapeutics so that the probability of such side-effects being experienced by patients can be minimized.

The combination of nanomedicine with light- based therapies and their applications in 3D platforms have enabled certain drawbacks associated with the use of conventional chemotherapy methods and simple 2D cultures to be addressed thus allowing more reliable data to be generated. This e-book highlights the breakthroughs made in this field and aims to inform the reader about the developments achieved in cancer therapy research using minimally invasive techniques such as Photodynamic Therapy and Photochemical Internalization alongside nano-delivery systems as well as their applications in various 3D cancer models. The reader will also be introduced to natural and synthetic materials used for the production of nanoparticles and the biological 3D platforms.

Layla Mohammad-Hadi and Marym Mohammad-Hadi
September 2020

Acknowledgments

We are grateful to God almighty for giving us the strength, knowledge, ability, and opportunity to produce this e-book. The authors also have great pleasure in acknowledging their gratitude to Professor Michael Hamblin for his invitation to write this book. We would like to dedicate this book to our parents for their endless love, care, and support throughout the years. The blessings of our parents have tremendously contributed toward helping us reach our goals in life. They have patiently held our hands through the years and motivated us to try and achieve our best in every situation. Finally, the authors would like to thank the production team at Morgan & Claypool and IOP Publishing for their help and guidance.

Layla Mohammad-Hadi and Marym Mohammad-Hadi
September 2020

CHAPTER 1

Introduction

1.1 INTRODUCTION

The therapeutic potential of several minimally invasive approaches have been investigated in different cancers in recent years as an attempt to find alternatives to surgery. Several ablative therapies such as radiofrequency ablation, laser interstitial ablation, focused ultrasound ablation, and cryotherapy function either through local freezing or heat to induce cellular death and tumor destruction have been proposed. However, such techniques are still under development and although safe, they have complications such as infection, bleeding and skin burns associated with them [1–5].

Other types of minimally invasive techniques for cancer treatment include energy-based therapies such as photodynamic therapy (PDT) and its modified counterpart treatment known as photochemical internalization (PCI) that can potentially be applied to cancers as with radiotherapy with possibly fewer side-effects.

As a treatment PDT is clinically applicable to different cancers as well as certain non-cancerous conditions [6]. PDT requires the administration of a light activatable photosensitizer which upon excitement, induces the production of reactive oxygen species (ROS) that cause damage to target cells.

PCI is a technique derived from PDT which enhances the delivery of therapeutic macromolecules and certain smaller molecules that are susceptible to endolysosomal entrapment and degradation into the cytosol through the use of an amphiphilic photosensitizer. The efficacy of this drug delivery technique has been proven in different *in vitro* monolayer and *in vivo* cancer models which also include drug-resistant cancers [7–10]. The potential of this technique in treating head and neck cancer has been recently demonstrated in a clinical trial study using bleomycin as the chemotherapeutic agent [11].

Until recently, conventional two-dimensional (2D) cell cultures where cells grow on uniform surfaces in monolayer forms and *in vivo* models were the main methods of carrying out cancer and drug discovery studies [12]. While the monolayer cultures have the advantages of easy preparation, maintenance, and monitoring [13, 14], they do not properly integrate the vital interactions between the cells and the surrounding extracellular matrix (ECM) that exist *in vivo* and consist mainly of structural proteins [15]. Furthermore, such models exhibit restricted cell-to-cell interaction since the main interaction occurs between the cells and the host surface which is normally plastic. The lack of such interactions in monolayer cultures may influence proliferation, signal transduction, and therapeutic response of cells by causing the adhesion properties and organization of cancer cells to differ from their *in vivo* counterparts [16, 17]. Moreover, the lack of ECM surrounding cells in monolayer cultures allows nanoparticles and drugs to reach the cells without encountering physical barriers while the *in vivo* models would present the nanoparticles with hinderance from the existing ECM [18]. Thus, exposing the cancer cells in 2D cultures to a constant environment which provides a steady supply of oxygen and nutrients precludes the model from imitating *in vivo* cancer tissues which normally experience nutrient and oxygen gradients [19] that could result in altered gene expression patterns. As an important component of the connective tissue in *in vivo* models the ECM, also known as the stroma when surrounding tumors allows cancer cells to grow and proliferate while in contact with it [20, 21]. Therefore, the 3D models which incorporate ECM materials are able to mimic cancer growth, structural features, and the 3D cellular organization which is normally observed *in vivo* [22–24]. The cell-stroma interactions displayed in 3D models can also affect factors such as therapeutic response [25], penetration of nanoparticles and drugs, anti-apoptotic signaling, multicellular resistance, and hypoxia [18, 26, 27].

In this book the applications of PDT and PCI in different biological 3D cancer models have been discussed. The incorporation of nanoparticles in PDT treatment, their mechanisms of action and advantages are also taken into consideration alongside the usability of various types of biological 3D models.

1.2 REFERENCES

[1] Vlastos, G. and Verkooijen, H. M. 2007. Minimally invasive approaches for diagnosis and treatment of early-stage breast cancer, *Oncologist*, 12:1–10. DOI: 10.1634/theoncologist.12-1-1. 1

[2] Friedman, M., Mikityansky, I., Kam, A., Libutti, S. K., Walther, M. M., Neeman, Z., Locklin, J. K., and Wood, B. J. 2004. Radiofrequency ablation of cancer, *Cardiovasc. Intervent. Radiol.*, 27:427–34. DOI: 10.1007/s00270-004-0062-0. 1

[3] Swartz, L. K., Holste, K. G., Kim, M. M., Morikawa, A., and Heth, J. 2019. Outcomes in patients treated with laser interstitial thermal therapy for primary brain cancer and brain metastases, *Oncologist*, 24:e1467–e1470. DOI: 10.1634/theoncologist.2019-0213. 1

[4] Mauri, G., Nicosia, L., Xu, Z., Di Pietro, S., Monfardini, L., Bonomo, G., Varano, G. M., Prada, F., Vigna, P. D., and Orsi, F. 2018. Focused ultrasound: tumour ablation and its potential to enhance immunological therapy to cancer, *Br. J. Radiol.*, 91:20170641. DOI: 10.1259/bjr.20170641. 1

[5] Theodorescu, D. 2004. Cancer cryotherapy: Evolution and biology, *Rev. Urol.*, 6:S9-S19. 1

[6] Dolmans, D. E., Fukumura, D., and Jain, R. K. 2003. Photodynamic therapy for cancer, *Nat. Rev. Cancer.*, 3:380–387. DOI: 10.1038/nrc1071. 1

[7] Bull-Hansen, B., Berstad, M. B., Berg, K., Cao, Y., Skarpen, E., Fremstedal, A. S., Rosenblum, M. G., Peng, Q., and Weyergang, A. 2015. Photochemical activation of MH3-B1/rGel: A HER2-targeted treatment approach for ovarian cancer, *Oncotarget.*, 6:12436–12451. DOI: 10.18632/oncotarget.3814. 1

[8] Berg, K., Weyergang, A., Prasmickaite, L., Bonsted, A., Hogset, A., Strand, M. T., Wagner, E., and Selbo, P. K. 2010. Photochemical internalization (PCI): A technology for drug delivery, *Meth. Mol. Biol.*, 635:133–145. DOI: 10.1007/978-1-60761-697-9_10. 1

[9] Martinez de Pinillos Bayona, A., Moore, C. M., Loizidou, M., MacRobert, A. J., and Woodhams, J. H. 2015. Enhancing the efficacy of cytotoxic agents for cancer therapy using photochemical internalisation, *Int. J. Cancer*, 138:1049–1057. DOI: 10.1002/ijc.29510. 1

[10] Weyergang, A., Berstad, M. E., Bull-Hansen, B., Olsen, C. E., Selbo, P. K., and Berg, K. 2015. Photochemical activation of drugs for the treatment of therapy-resistant cancers, *Photochem. Photobiol. Sci.*, 14:1465–1475. DOI: 10.1039/c5pp00029g. 1

[11] Sultan, A. A., Jerjes, W., Berg, K., Hogset, A., Mosse, C. A., Hamoudi, R., Hamdoon, Z., Simeon, C., Carnell, D., Forster, M., and Hopper, C. 2016. Disulfonated tetraphenyl chlorin (TPCS2a)—induced photochemical internalisation of bleomycin in patients with solid malignancies: A phase 1, dose-escalation, first-in-man trial, *Lancet Oncol.*, 17:1217–1229. DOI: 10.1016/s1470-2045(16)30224-8. 1

[12] Nyga, A., Loizidou, M., Emberton, M., and Cheema, U. 2013. A novel tissue engineered three-dimensional in vitro colorectal cancer model, *Acta Biomater.*, 9:7917–7926. DOI: 10.1016/j.actbio.2013.04.028. 1

[13] Hsieh, C., Chen, Y., Huang, S., Wang, H., and Wu, M. 2015. The effect of primary cancer cell culture models on the results of drug chemosensitivity assays: The application of perfusion microbioreactor system as cell culture vessel, *BioMed Res. Int.*, pages 1–10. DOI: 10.1155/2015/470283. 1

[14] Ma, H. L., Qiao, J., Han, S., Wu, Y., Cui Tomshine, J., Wang, D., Gan, Y., Zou, G., and Liang, X. J. 2012. Multicellular tumor spheroids as an in vivo-like tumor model for threedimensional imaging of chemotherapeutic and nano material cellular penetration, *Mol. Imag.*, 11:487–498. DOI: 10.2310/7290.2012.00012. 1

[15] Alemany-Ribes, M., Garcia-Diaz, M., Busom, M., Nonell, S., and Semino, C. E. 2013. Toward a 3D cellular model for studying in vitro the outcome of photodynamic treat-

ments: Accounting for the effects of tissue complexity, *Tissue Eng. Part A*, 19:1665–1674. DOI: 10.1089/ten.tea.2012.0661. 1

[16] Zanoni, M., Piccinini, F., Arienti, C., Zamagni, A., Santi, S., Polico, R., Bevilacqua, A., and Tesei, A. 2016. 3D tumor spheroid models for in vitro therapeutic screening: A systematic approach to enhance the biological relevance of data obtained, *Sci. Rep.*, 6:19103. DOI: 10.1038/srep19103. 2

[17] Zhang, M., Boughton, P., Rose, B., Lee, C., and Hong, A. 2013. The use of porous scaffold as a tumor model, *Int. J. Biomater.*, pages 1–9. DOI: 10.1155/2013/396056. 2

[18] Xu, X., Sabanayagam, C. R., Harrington, D. A., Farach-Carson, M. C., and Jia, X. 2014. A hydrogel-based tumor model for the evaluation of nanoparticle-based cancer therapeutics, *Biomaterials*, 35:3319–3330. DOI: 10.1016/j.biomaterials.2013.12.080. 2

[19] Keith, B. and Simon, M. C. 2007. Hypoxia-inducible factors, stem cells, and cancer, *Cell*, 129:465–472. DOI: 10.1016/j.cell.2007.04.019. 2

[20] Edmondson, R., Broglie, J., Adcock, A., and Yang, L. 2014. Three-dimensional cell culture systems and their applications in drug discovery and cell-based biosensors, *ASSAY Drug Devel. Technol.*, 12:207–218. DOI: 10.1089/adt.2014.573. 2

[21] Yamada, K. M. and Cukierman, E. 2007. Modeling tissue morphogenesis and cancer in 3D, *Cell*, 130:601–610. DOI: 10.1016/j.cell.2007.08.006. 2

[22] Antoni, D., Burckel, H., Josset, E., and Noel, G. 2015. Three-dimensional cell culture: A breakthrough in vivo, *Int. J. Mol. Sci.*, 16:5517–5527. DOI: 10.3390/ijms16035517. 2

[23] van Duinen, V., Trietsch, S. J., Joore, J., Vulto, P., and Hankemeier, T. 2015. Microfluidic 3D cell culture: From tools to tissue models, *Curr. Opin. Biotechnol.*, 35:118–126. DOI: 10.1016/j.copbio.2015.05.002. 2

[24] Huh, D., Hamilton, G. A., and Ingber, D. E. 2011. From 3D cell culture to organs-on-chips, *Trends Cell Biol.*, 21:745–754. DOI: 10.1016/j.tcb.2011.09.005. 2

[25] Shin, C. S., Kwak, B., Han, B., and Park, K. 2013. Development of an in vitro 3D tumor model to study therapeutic efficiency of an anticancer drug, *Mol. Pharm.*, 10:2167–2175. DOI: 10.1021/mp300595a. 2

[26] Celli, J. P., Rizvi, I., Blanden, A. R., Massodi, I., Glidden, M. D., Pogue, B. W., and Hasan, T. 2014. An imaging-based platform for high-content, quantitative evaluation of therapeutic response in 3D tumour models, *Sci. Rep.*, 4:3751. DOI: 10.1038/srep03751. 2

[27] Zhao, J., Lu, H., Wong, S., Lu, M., Xiaob, P., and Stenze, M. H. 2017. Influence of nanoparticle shapes on cellular uptake of paclitaxel loaded nanoparticles in 2D and 3D cancer models, *Polym. Chem.*, 8:3317–3326. DOI: 10.1039/c7py00385d. 2

CHAPTER 2

Photodynamic Therapy and Photochemical Internalization

2.1 INTRODUCTION

Photodynamic Therapy (PDT) and Photochemical Internalization (PCI) are both minimally invasive, light-based techniques used for the treatment of cancers. PDT has been clinically approved for treating various cancers and conditions using certain photosensitizers and has many benefits. PCI is a modification of PDT that bears some potential advantages such as reduction in photosensitivity and lack of multi-drug resistance over this already clinically used treatment. This section explores (1) the mechanisms of PDT and PCI, (2) pre-clinical and clinical applications of PDT and PCI, and (3) advantages and disadvantages of PDT and PCI.

2.2 PHOTODYNAMIC THERAPY

PDT has been established as a minimally invasive technique for treating different forms of cancer as well as non-malignant lesions [1, 2]. PDT functions through the activation of an administered photosensitizer using light of a specific wavelength (e.g., blue, red, near infrared (NIR) light) which results in the generation of cytotoxic reaction oxygen species (ROS) from molecular oxygen. The solid cancer types which PDT has been clinically approved to treat include prostate [3], head and neck, skin and esophagus [4]. This technique has also been employed to treat non-cancerous conditions such as age-related macular degeneration, atherosclerosis, and bacterial infections [5].

The advantages associated with using PDT include greatly reducing the requirement of major surgery, shortening the duration of recovery, promoting good healing, maintaining the integrity and function of organs and holding a low risk of local and systemic treatment-related morbidity as well as side-effects [6–9]. Another benefit of this technique is its potential for repeated applications as well as use after different treatments such as surgery, chemotherapy, and radiotherapy without stimulating any immunosuppressive or myelosuppressive effects [2, 10]. PDT allows light delivery to be targeted to the lesions. The lights used for clinical applications generally are red or NIR since tissue has more transparency in this wavelength range beyond strong hemoglobin absorption. Very superficial regions can, however, be treated using blue light excitation. Tumor-selective treatment with PDT can be limited due to insufficient selective uptake of the photosensitizer post intravenous administration. However, in comparison to radio-

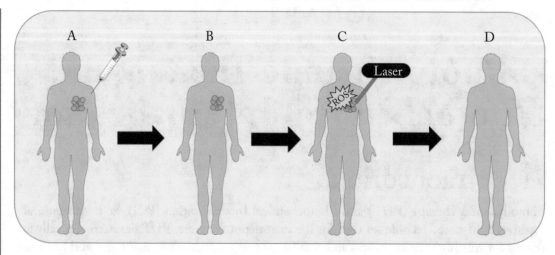

Figure 2.1: Application of PDT in patients. (a) Administration of photosensitizer. (b) Accumulation of photosensitizer in the tumor. (c) PDT treatment using a laser and the production of ROS. (d) Eradication of the tumor and recovery of the patient.

therapy which causes damage to normal tissue, PDT offers the therapeutic benefit of promoting good healing of normal tissue adjacent to the tumor.

The PDT mechanism of action and the mode of cell death induced by this treatment depends on several factors, e.g., photosensitizer localization, genotype of cells, and PDT dosimetry (e.g., light intensity) [11, 12]. Since most photosensitizers do not have the tendency to accumulate within the nuclei [13], cells are not susceptible to DNA related damage, mutations, or carcinogenesis [12]. Photosensitizers that localize mostly within the mitochondria tend to induce apoptosis [14], whereas photosensitizers that localize predominantly within the plasma membrane mainly stimulate necrosis following exposure to light [15]. The shifting from apoptotic to necrotic mode of cell death occurs with increase in damage to the cell which leads to swift cell lysis rather than an orderly programmed form of cell death [16]. Photodamage could also trigger a cytoprotective response called autophagy [17].

In their lowest energy level, photosensitizers exist in a stable singlet state (i.e., no net electronic spin) [18]. However, they are raised to an excited and short-lived singlet state through the absorption of a light photon of a certain wavelength [19, 20]. The photosensitizer may then return to ground state through internal conversion by either losing its energy as heat or emitting a photon as fluorescence which can be used clinically for photodetection purposes [21]. However, to produce a therapeutic photodynamic effect, the photosensitizer, must be converted to the relatively long-lived triplet state via a change in its electronic spin. This process is known as intersystem crossing [6]. Most photosensitizers have > 0.5 probability of converting to the triplet state.

In the presence of an adequate oxygen supply, the triplet excited photosensitizer can undergo either a type I or type II reaction [22]. Type I reaction involves a direct reaction between the photosensitizer and a substrate through proton or electron transfer to generate radicals, which then interact with oxygen molecules to produce oxygenated products or ROS that can cause damage to the integrity of the cell membrane [23]. However, in a type II reaction, the energy of the sensitizer is transferred directly to the oxygen molecules to form singlet oxygen species, which are recognized as highly potent ROS in PDT [24].

The reactive nature of the singlet oxygen and its ability to diffuse only around 0.01–0.02 μm in its short lifetime allows the reaction to occur within a limited distance and volume leading to a localized cytotoxic response being created [25].

2.3   LIGHT SOURCES USED IN PDT

A range of laser and non-laser light sources can be employed for photodynamic therapy [26]. Wide spectrum light sources such as arc lamps are among the cost-effective and convenient-to-use sources that can be utilized to activate photosensitizers [27, 28]. The drawback associated with the use of these lamps includes the difficulty behind coupling them to light delivery fibers without needing to reduce their optical power, calculating the amount of the efficient light dose delivered as well as the limited power output (1 W maximum) and the requirement of filters to eradicate UV radiation in addition to infrared emission which can cause heating [25].

The discovery of lasers has helped to overcome many issues related to wide spectrum light sources. The ability of the lasers to emit focused light beams with the precise required wavelengths has proven to be a major discovery for PDT [29, 30]. The developmental progress in the semiconductor diode technology has resulted in the production of more portable and cost-effective systems with high power output and compact structures [31, 32]. Most of these systems contain an internal unit for dosimetric calculations as well as built-in treatment programs making them more user friendly [28, 33]. The one disadvantage of these laser diodes is that they provide only one output wavelength [25]. It is possible to use light-emitting diodes (LEDs) in a clinical setting [34] as they are small in size and more cost effective than the above-mentioned light sources and have the capability to provide maximum power output of 150 mW/cm^2 at wavelengths ranging from 350–1100 nm [20].

Optical fiber technology has been instrumental in delivering light to the treatment site to activate the photosensitizer in recent years [35]. Optical fibers take advantage of their ability to deliver light with a higher accuracy and homogenous distribution to target sites to illuminate at different localizations [36]. In order to treat superficial cancers, optic fibers containing a lens on their tip are employed to distribute the light across the targeted site [28]. However, in the case of treating hollow organs, e.g., bladder and esophagus, cylindrical diffusers coupled with inflated balloons are mainly used as illumination tools to provide a uniform light distribution [37–39]. Furthermore, healthy tissue areas can be protected from illumination by a black coating on a side of the balloon [25].

2.4   CLINICAL TRIALS OF PDT IN CANCER PATIENTS

PDT can be applied clinically as a monotherapy or after surgery, chemotherapy, and radio-therapy. This treatment has been tested in different trials worldwide and has obtained approval for treatment of several solid cancers, e.g., prostate, head and neck. The trials conducted include treatment of advanced head and neck cancer, gastrointestinal cancers (esophagus, stomach, pancreas), pulmonary cancers, and ovarian cancer as well as pre-cancerous lesions such as actinic keratoses to prevent its progression to squamous cell carcinoma [3, 40–42].

Breast cancer in patients has also been a focus of several PDT trials including one study by Li et al. (2011) which combined PDT with immunotherapy to treat patients with metastatic breast cancer (stage 3 or 4). From all the patients that participated in this study, one showed a complete response, four demonstrated partial responses, two maintained the disease in stable state, and two patients showed disease progression. Moreover, the study found no serious side effects or deaths associated with this treatment [2, 43]. In another clinical trial study which employed Photofrin as photosensitizer for the PDT treatment of ovarian cancer in 13 patients, no recurrence of the disease was experienced by 2 out of 13 patients [42].

In a different study, Hahn et al. (2006) conducted a phase II trial on the effect of PDT on 33 patients with ovarian cancer also using photofrin as a photosensitizer. While the results of this study indicated an overall survival of 20 months, the treatment was not found to result in significant complete responses or long-term tumor control in the patients [44].

A phase l/lla trial carried out by Banerjee et al. (2017) to investigate the effect of verteporfin (BPD)—PDT on 11 patients with primary breast cancer found that in all patients a plateau with no expansion in the diameter of the necrotic area was achieved by increasing the light dose. The light doses used in this study ranged from 20–50 J, however no information was provided regarding the minimum required light dose for the treatment as the trial is still ongoing [45].

The effectiveness and safety of Padeliporfin vascular-targeted PDT has been compared to active surveillance in men diagnosed with low-risk, localized prostate cancer in a phase III randomized controlled clinical trial. Patients with no previous treatment were selected and randomly assigned to vascular-targeted PDT or active surveillance. The vascular-targeted PDT involved intravenous administration of padeliporfin (4 mg/kg) over 10 min and activation with laser light 753 nm (fixed power 150 mW/cm for 22 min and 15 sec) through insertion of optical fibers. The active surveillance, however, involved biopsy at every 12 months and prostate-specific antigen measurement as well as digital rectal examination at every 3 months. In the vascular-targeted PDT group 58/206 (28%) of the patients showed disease progression at 24 months while 120/207 (58%) patients experienced such progression in the active surveillance group. Furthermore, 49% of patients had negative prostate biopsy results 24 months post-treatment with vascular-targeted PDT whereas such finding was present in only 14% of patients in the active surveillance group. The most common serious adverse event was urine retention (15 patients) in the vascular-targeted PDT group and myocardial infarction (3 patients) in the active surveillance group. Therefore, the padeliporfin vascular targeted PDT was found to be an effective and safe tissue-preserving treatment for low-risk, localized prostate cancer [3].

2.5   CLINICAL TRIALS IN NON-CANCEROUS DISEASES

Photodynamic therapy has been used in clinical trials to treat non-cancerous conditions such as periodontitis, actinic keratoses (AK), and age-related macular degeneration.

Christodoulidas et al. (2008) randomly assigned 24 patients with severe periodontitis to either scaling and root planing followed by a single episode of PDT treatment group (test) or scaling and root planing alone group (control). No additional improvement was observed in probing depth (PD), gingival recession, and clinical attachment level (CAL) following application of PDT after scaling and root planning at 3 and 6 months post treatment. However, the bleeding score of the patients was significantly reduced in the test group compared to the control group [46].

A similar study was carried out by Segarra-Vidal et al. (2017) where 20 healthy controls and 40 patients with periodontitis were enrolled with the patients being randomly assigned to either the scaling and root planing (SRP) treatment group or SRP + PDT group. This study also did not find any additional improvement in the PD and CAL parameters following SRP + PDT compared to SRP alone. However, the presence of bacterium A. actinomycetemcomitans was significantly reduced [47].

A phase-IV clinical trial study examined the efficacy and safety of aminolaevulinic acid-based photodynamic therapy (ALA-PDT) in patients with 6–12 AK lesions either on their face or scalp. The populations which experienced clearing of the target AK lesions post a single ALA-PDT treatment were 76% and 72%, respectively. The population of patients that received a second ALA-PDT treatment, for the target AKs that were still present at month 2 was 60%. The rate of recurrence of histologically confirmed AKs was found to be 19% [48].

Treatment of Classic Choroidal Neovascularization (CNV) in age-related macular degeneration using ranibizumab vs. PDT with verteporfin has been studied in a phase III clinical trial. The 2-year study showed that treatment with ranibizumab lead to greater clinical benefit than verteporfin PDT in the patients. Furthermore, the rates of serious adverse events were found to be low [49].

2.6   PHOTOCHEMICAL INTERNALIZATION

As a commonly used treatment for numerous metastatic cancers, chemotherapy is accompanied by side effects and the risk of developing multidrug resistance [50]. Enhancement of drug delivery and the prevention of drug resistance development are therefore essential for improving treatment response. One factor that lowers the efficacy of immunotoxic and biologic agents is their uptake through endocytosis due to the large sizes of the agents. This endocytic uptake acts as a barrier against the intracellular targeting of components by the therapeutic macromolecules [51] as the agents are unable to reach their target efficiently through cytosolic diffusion. The entrapment of the agents within endosomes and subsequently lysosomes also makes them prone to degradation by proteolytic enzymes which further limits their efficacy and bioavailability. In some cases even for small molecules the poor bioavailability due to endocytic uptake can also be attributed to low water solubility and/or reduced penetration through cell membrane [52].

Photochemical Internalization (PCI) uses low dose PDT to assist with the cytosolic delivery of therapeutic macromolecules and certain small drugs, which are susceptible to endolysosomal entrapment and degradation to their cytosolic sites of action [53–56]. In the same way as PDT, PCI is spatially selective as the cytosolic release is triggered by light delivered at the target site. PCI makes use of an amphiphilic photosensitizer that localizes in the membranes of endolysosomal vesicles, while the cytotoxic agent resides within the vesicles. Upon photoactivation, the photosensitized endolysosomal membranes become damaged by ROS thus releasing the entrapped bioactive agent into the cytosol allowing them to reach their intracellular target [57, 58].

Endolysosomal pH-responsive drug release systems, redox-responsive drug release systems and lysosomal enzyme-responsive drug release systems are other forms of cytosolic drug delivery [59]. However, unlike PCI these systems lack spatial selectivity and cannot control the timing of the drug release or the possible toxicity triggered by the delivery system.

2.7   DIFFERENT MECHANISMS OF ENDOCYTOSIS

Endocytosis occurs through various mechanisms. The most common pathways include clathrin-mediated endocytosis, caveolin-mediated endocytosis, and clathrin- and caveolin-independent endocytosis. Clathrin-mediated endocytosis involves the binding of cell surface receptors to appropriate macromolecular ligands which then accumulate within specific regions of the plasma membrane called coated pits that contain clathrin. These pits then pinch off and form coated vesicles. Upon entering the cell, the internalized macromolecules become trapped within endolysosomal vesicles and ultimately undergo degradation by the lysosomal enzymes. Caveolae-dependent endocytosis on the other hand is a clathrin-independent endocytosis process during which bulb shaped plasma membrane invaginations known as caveolae are formed. In this type of endocytosis, the caveolae pinch off the plasma membrane and while some of the caveolae try to fuse back with the plasma membrane most reach early endosome and are then recycled back to the plasma membrane [60, 61].

2.8   CLINICAL TRIALS OF PCI

PCI has been tested in patients with advanced and recurrent solid malignancies in a phase 1 clinical trial. In 2016, Sultan et al. examined safety and tolerability of photosensitizer disulfonated tetraphenyl chlorin (TPCS$_{2a}$) also known as Fimaporfin or Amphinex, in mediating PCI of chemotherapy drug bleomycin in 22, 18–84-year-old patients with local recurrent, advanced,

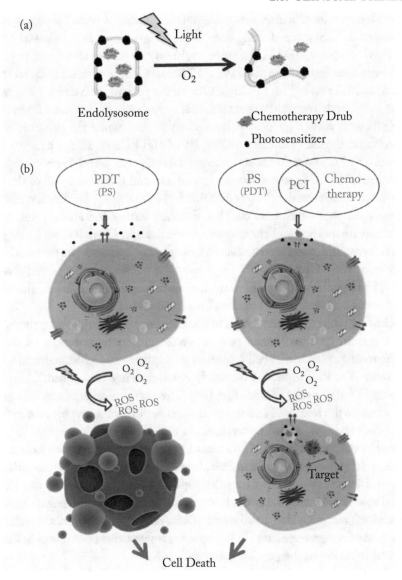

Figure 2.2: Comparison of photochemical internalization (PCI) and photodynamic therapy (PDT). (a) In PCI the photosensitizer and chemotherapeutic agent are taken up by the cancer cells through endocytosis and released into the cytosol to reach their intracellular targets via photo-induced rupture of the endolysosomal membrane following prolonged irradiation. (b) The use of low dose PDT to release endolysosomally entrapped agents in PCI allows the cells to incur sublethal photooxidative damage. Since PCI is site-specific, the photosensitizers used for this treatment technique do not induce direct cell death like those employed in PDT. Reprinted with permission from [54]. Copyright 2015 by John Wiley & Sons, Inc.

or metastatic cutaneous or subcutaneous malignancies. Patients were subjected to slow intravenous injection of $TPCS_{2a}$ at a starting dose of 0.25 mg/kg on day 0 followed by intravenous infusion of a fixed dose of 15,000 IU/m^2 bleomycin on day 4. The surface of target tumor was exposed to 652-nm laser light (fixed at 60 J/cm^2) 3 hr later. Three patients also received increased doses (0.5, 1.0, and 1.5 mg/kg) of $TPCS_{2a}$. Out of the patients recruited, 12 completed the 3 months follow-up period. The results reported local or systemic PCI-related adverse events. Local adverse effects were caused by local inflammatory process while the systemic adverse events mainly resulted from the skin photosensitizing effect of $TPCS_{2a}$. In the patients that received the higher doses of $TPCS_{2a}$, the maximum tolerated dose was found to be 1.0 mg/kg. Overall, the administration of $TPCS_{2a}$ was discovered to be safe and tolerable by all of the trial patients with recommended treatment dose being identified as 0.25 mg/kg for future trials [62].

A more recent phase 1 clinical trial has finished testing the efficacy and safety of PCI using Amphinex (Fimaporfin) and chemotherapy medication Gemcitabine in 16 patients with Cholangiocarcinoma. The study was conducted in 4 dose cohorts with the patients being exposed to up to 8 cycles of the PCI treatment. The two highest dose cohorts caused more than 20% reduction in 17/19 target lesions with 12 of the lesions becoming untraceable. The study reported no unexpected serious adverse effects or dose limiting toxicities [63].

PCI technology has also been explored for the purpose of improving therapeutic cancer vaccination. A current trial by Hogset et al. is investigating the potential of the vaccine (fimaVacc) in enhancing cytotoxic T-cell responses to peptide- and protein-based vaccination in healthy volunteers. The PCI aspect of the study involves using Fimaporfin (Amphinex) and a toll-like receptor (TLR) agonist formulation [64]. The pre-clinical study by this group showed that PCI can promote the release of endocytosed short peptide antigens into the cytosol of Antigen presenting cells (APCs). Such improvement in the delivery of the peptide antigen lead to a 30-fold increase in the formation of MHC class I/peptide complex as well as surface presentation. A 30–100-fold enhancement in the activation of antigen-specific cytotoxic T cells (CTL) was also observed in comparison to using the peptide alone. The dose of the PCI treatment played a crucial role in achieving an optimal effect and stimulating the maturation of immature dendritic cells which also provided an adjuvant effect. An *in vivo* study also confirmed the successful stimulation of antigen-specific CTL responses toward cancer antigens in C57BL/6 mice after intradermal injection of the peptide using PCI [65].

2.9   POTENTIAL ADVANTAGES OF PCI OVER PDT

Despite sharing some common features, PCI has numerous potential benefits over PDT. First, PCI relies on sub-lethal PDT to cause cytosolic release of endolysosomally trapped agents. This could result in diminished skin photosensitivity as a lower dose of the sensitizer is needed for PCI. Second, in comparison to PDT, PCI may reduce and delay injury to endothelial cells, which could possibly be due to the utilization of lower light fluences. This may decrease the probability of vascular shutdown and therefore prevent tumor oxygenation which is important for generation of singlet oxygen from being significantly affected by PCI, thus leading to greater tumor eradication [66–68]. Third, photosensitizers used in PCI, e.g., TPPS$_{2a}$, are not substrates for efflux via ABCG2 transporters which are known to mediate cellular multidrug resistance. This is an important advantage since studies have shown that many of the second generation photosensitizers used for PDT are subject to efflux by ABCG2 [55, 69–72].

Lou et al. (2006) were able to overcome doxorubicin resistance in MCF-7/ADR cells using PCI thus making the sensitivity of these cells toward the drug comparable to that seen in MCF-7 cells. Weak-base drugs such as doxorubicin are believed to be predisposed to endosomal entrapment in multiple drug resistant cells as these cells display increased acidification of the endocytic vesicles as well as elevated cytosolic pH levels [73]. In another study by Bostad et al. (2013), PCI was employed for the targeting of CD133-positive colon cells (WiDr) using an immunotoxin consisting of mAb CD133/1 bound to ribosome inactivating plant toxin saporin (anti-CD133/1-sap). The results showed that PCI of anti-CD133/1-sap enhanced the effect of the immunotoxin. Furthermore, using PCI allowed the circumvention of PDT resistance by the WiDr cells [74].

2.10 REFERENCES

[1] Dolmans, D. E., Fukumura, D., and Jain, R. K. 2003. Photodynamic therapy for cancer, *Nat. Rev. Cancer*, 3:380–387. DOI: 10.1038/nrc1071. 7

[2] Banerjee, S. M., MacRobert, A. J., Mosse, C. A., Periera, B., Bown, S. G., and Keshtgar, M. R. S. 2017. Photodynamic therapy: Inception to application in breast cancer, *Breast*, 31:105–113. DOI: 10.1016/j.breast.2016.09.016. 7, 10

[3] Azzouzi, A. R., Vincendeau, S., Barret, E., Cicco, A., Kleinclauss, F., van der Poel, H. G., Stief, C. G., Rassweiler, J., Salomon, G., Solsona, E., Alcaraz, A., Tammela, T. T., Rosario, D. J., Gomez-Veiga, F., Ahlgren, G., Benaghou, F., Gaillac, B., Amzal, B., Debruyne, M. J., Fromont, G., Gratzke, C., Emberton, M., and PCM301 Study Group. 2017. Padeliporfin vascular-targeted photodynamic therapy versus active surveillance in men with low-risk prostate cancer (CLIN1001 PCM301): An open-label, phase 3, randomised controlled trial, *Lancet Oncol.*, 18:181–191. DOI: 10.1016/s1470-2045(16)30661-1. 7, 10, 11

[4] Mallidi, S., Anbil, S., Bulin, A. L., Obaid, G., Ichikawa, M., and Hasan, T. 2016. Beyond the barriers of light penetration: Strategies, perspectives and possibilities for photodynamic therapy, *Theranostics*, 6:458–487. DOI: 10.7150/thno.16183. 7

[5] Abrahamse, H. and Hamblin, M. R. 2016. New photosensitizers for photodynamic therapy, *Biochem. J.*, 473:347–364. DOI: 10.1042/bj20150942. 7

[6] Agostinis, P., Berg, K., Cengel, K. A., Foster, T. H., Girotti, A. W., Gollnick, S. O., Hahn, S. M., Hamblin, M. R., Juzeniene, A., Kessel, D., Korbelik, M., Moan, J., Mroz,

P., Nowis, D., Piette, J., Wilson, B. C., and Golab, J. 2011. Photodynamic therapy of cancer: An update, *CA Cancer J. Clin.*, 61:250–281. DOI: 10.3322/caac.20114. 7, 8

[7] Hopper, C. 2000. Photodynamic therapy: A clinical reality in the treatment of cancer, *Lancet Oncol.*, 1:212–219. DOI: 10.1016/s1470-2045(00)00166-2. 7

[8] Nyst, H., Tan, I., Stewart, F., and Balm, A. 2009. Is photodynamic therapy a good alternative to surgery and radiotherapy in the treatment of head and neck cancer?, *Photodiagn. Photodyn. Ther.*, 6:3–11. DOI: 10.1016/j.pdpdt.2009.03.002. 7

[9] Brown, S. B., Brown, E. A., and Walker, I. 2004. The present and future role of photodynamic therapy in cancer treatment, *Lancet Oncol.*, 5:497–508. DOI: 10.1016/s1470-2045(04)01529-3. 7

[10] Misra, R., Acharya, S., and Sahoo, S. K. 2010. Cancer nanotechnology: Application of nanotechnology in cancer therapy, *Drug Discov. Today*, 15:842–850. DOI: 10.1016/j.drudis.2010.08.006. 7

[11] Acedo, P., Stockert, J. C., Canete, M., and Villanueva, A. 2014. Two combined photosensitizers: A goal for more effective photodynamic therapy of cancer, *Cell Death Dis.*, 5:e1122. DOI: 10.1038/cddis.2014.77. 8

[12] O'Connor, A. E., Gallagher, W. M., and Byrne, A. T. 2009. Porphyrin and nonporphyrin photosensitizers in oncology: preclinical and clinical advances in photodynamic therapy, *Photochem Photobiol.*, 85:1053–1074. DOI: 10.1111/j.1751-1097.2009.00585.x. 8

[13] Slastnikova, T. A., Rosenkranz, A. A., Lupanova, T. N., Gulak, P. V., Gnuchev, N. V., and Sobolev, A. S. 2012. Study of efficiency of the modular nanotransporter for targeted delivery of photosensitizers to melanoma cell nuclei in vivo, *Dokl. Biochem. Biophys.*, 446:235–237. DOI: 10.1134/s1607672912050146. 8

[14] Kessel, D. 2015. Apoptosis and associated phenomena as a determinants of the efficacy of photodynamic therapy, *Photochem. Photobiol. Sci.*, 14:1397–1402. DOI: 10.1039/c4pp00413b. 8

[15] Kessel, D. and Oleinick, N. L. 2010. Photodynamic therapy and cell death pathways, *Meth. Mol. Biol.*, 635:35–46. DOI: 10.1007/978-1-60761-697-9_3. 8

[16] Marchal, S., Fadloun, A., Maugain, E., D'Hallewin, M., Guillemin, F., and Bezdetnaya, L. 2005. Necrotic and apoptotic features of cell death in response to Foscan® photosensitization of HT29 monolayer and multicell spheroids, *Biochemical. Pharmacol.*, 69:1167–1176. DOI: 10.1016/j.bcp.2005.01.021. 8

[17] Kessel, D. 2015. Autophagic death probed by photodynamic therapy, *Autophagy*, 11:1941–1943. DOI: 10.1080/15548627.2015.1078960. 8

[18] Ion, R. M. 2008. ChemInform abstract: Photodynamic therapy (PDT): A photochemical concept with medical applications, *ChemInform*, 39. DOI: 10.1002/chin.200849276. 8

[19] Juarranz, A., Jaen, P., Sanz-Rodriguez, F., Cuevas, J., and Gonzalez, S. 2008. Photodynamic therapy of cancer. Basic principles and applications, *Clin. Trans. Oncol.*, 10:148–154. DOI: 10.1007/s12094-008-0172-2. 8

[20] Hempstead, J., Jones, D. P., Ziouche, A., Cramer, G. M., Rizvi, I., Arnason, S., Hasan, T., and Celli, J. P. 2015. Low-cost photodynamic therapy devices for global health settings: Characterization of battery-powered LED performance and smartphone imaging in 3D tumor models, *Sci. Rep.*, 5:10093. DOI: 10.1038/srep10093. 8, 9

[21] Kushibiki, T., Tajiri, T., Tomioka, Y., and Awazu, K. 2010. Photodynamic therapy induces interleukin secretion from dendritic cells, *Int. J. Clin. Exp. Med.*, 3:110–114. 8

[22] Ding, H., Yu, H., Dong, Y., Tian, R., Huang, G., Boothman, D. A., Sumer, B. D., and Gao, J. 2011. Photoactivation switch from type II to type I reactions by electron-rich micelles for improved photodynamic therapy of cancer cells under hypoxia, *J. Control Release*, 156:276–280. DOI: 10.1016/j.jconrel.2011.08.019. 9

[23] Rajesh, S., Koshi, E., Philip, K., and Mohan, A. 2011. Antimicrobial photodynamic therapy: An overview, *J. Indian Soc. Periodontol.*, 15:323–327. DOI: 10.4103/0972-124x.92563. 9

[24] Josefsen, L. B. and Boyle, R. W. 2008. Photodynamic therapy and the development of metal-based photosensitisers, *Metal-Based Drugs*, pages 1–23. DOI: 10.1155/2008/276109. 9

[25] Triesscheijn, M., Baas, P., Schellens, J. H., and Stewart, F. A. 2006. Photodynamic therapy in oncology, *Oncologist*, 11:1034–1044. DOI: 10.1634/theoncologist.11-9-1034. 9

[26] Valentine, R. M., Wood, K., Brown, C. T., Ibbotson, S. H., and Moseley, H. 2012. Monte Carlo simulations for optimal light delivery in photodynamic therapy of non-melanoma skin cancer, *Phys. Med. Biol.*, 57:6327–6345. DOI: 10.1088/0031-9155/57/20/6327. 9

[27] Chester, A., Martellucci, S., and Verga Scheggi, A. 2012. *Laser Systems for Photobiology and Photomedicine*, New York, Plenum Press. DOI: 10.1007/978-1-4684-7287-5. 9

[28] Huang, Z., Xu, H., Meyers, A. D., Musani, A. I., Wang, L., Tagg, R., Barqawi, A. B., and Chen, Y. K. 2008. Photodynamic therapy for treatment of solid tumors—potential and technical challenges, *Technol. Cancer Res. Treat.*, 7:309–320. DOI: 10.1177/153303460800700405. 9

[29] Porteous, M. S. and Rowe, D. J. 2014. Adjunctive use of the diode laser in non-surgical periodontal therapy: Exploring the controversy, *J. Dent. Hyg.*, 88:78–86. 9

[30] Santosa, V. and Limantara, L. 2008. Photodynamic therapy: New light in medicine world, *Indo. J. Chem.*, 8:279–291. DOI: 10.22146/ijc.21638. 9

[31] Hamblin, M. R. and Mroz, P. 2008. *Advances in Photodynamic Therapy*, Boston, Artech House. DOI: 10.1364/opn.7.7.000016. 9

[32] Mallidi, S., Mai, Z., Rizvi, I., Hempstead, J., Arnason, S., Celli, J., and Hasan, T. 2015. In vivo evaluation of battery-operated light-emitting diode-based photodynamic therapy efficacy using tumor volume and biomarker expression as endpoints, *J. Biomed. Optics*, 20:048003. DOI: 10.1117/1.jbo.20.4.048003. 9

[33] Kinhikar, R., Chaudhari, S., Kadam, S., Dhote, D., and Deshpande, D. 2012. Dosimetric validation of new semiconductor diode dosimetry system for intensity modulated radiotherapy, *J. Cancer Res. Ther.*, 8:86–90. DOI: 10.4103/0973-1482.95180. 9

[34] Opel, D. R., Hagstrom, E., Pace, A. K., Sisto, K., Hirano-Ali, S. A., Desai, S., and Swan, J. 2015. Light-emitting diodes: A brief review and clinical experience, *J. Clin. Aesthet. Dermatol.*, 8:36–44. 9

[35] Parker, S. 2013. The use of diffuse laser photonic energy and indocyanine green photosensitiser as an adjunct to periodontal therapy, *BDJ*, 215:167–171. DOI: 10.1038/sj.bdj.2013.790. 9

[36] Mohammad-Hadi, L., MacRobert, A. J., Loizidou, M., and Yaghini, E. 2018. Photodynamic therapy in 3D cancer models and the utilisation of nanodelivery systems, *Nanoscale*, 10:1570–1581. DOI: 10.1039/c7nr07739d. 9

[37] Barret, E. and Durand, M. 2015. *Technical Aspects of Focal Therapy in Localized Prostate Cancer*, 1st ed., pages 1–246, Paris, Springer. DOI: 10.1007/978-2-8178-0484-2. 9

[38] Shafirstein, G., Battoo, A., Harris, K., Baumann, H., Gollnick, S. O., Lindenmann, J., and Nwogu, C. E. 2016. Photodynamic therapy of non-small cell lung cancer. Narrative review and future directions, *Ann. Am. Thorac. Soc.*, 13:265–275. DOI: 10.1513/annalsats.201509-650fr. 9

[39] Simone, C. B. and Cengel, K. A. 2014. Definitive surgery and intraoperative photodynamic therapy: a prospective study of local control and survival for patients with pleural dissemination of non-small cell lung cancer, *Proc. SPIE Int. Soc. Opt. Eng.*, 8931:89310Y. DOI: 10.1117/12.2046679. 9

[40] Huang, Z. 2005. A review of progress in clinical photodynamic therapy, *Technol. Cancer Res. Treat.*, 4:283–293. DOI: 10.1177/153303460500400308. 10

[41] Oliveira, J., Monteiro, E., Santos, J., Silva, J. D., Almeida, L., and Santos, L. L. 2017. A first in human study using photodynamic therapy with Redaporfin in advanced head and neck cancer, *J. Clin. Oncol.*, 35. DOI: 10.1200/jco.2017.35.15_suppl.e14056. 10

[42] Wilson, J. J., Jones, H., Burock, M., Smith, D., Fraker, D. L., Metz, J., Glatstein, E., and Hahn, S. M. 2004. Patterns of recurrence in patients treated with photodynamic therapy for intraperitoneal carcinomatosis and sarcomatosis, *Int. J. Oncol.*, 24:711–717. DOI: 10.3892/ijo.24.3.711. 10

[43] Li, X., Ferrel, G. L., Guerra, M. C., Hode, T., Lunn, J. A., Adalsteinsson, O., Nordquist, R. E., Liu, H., and Chen, W. R. 2011. Preliminary safety and efficacy results of laser immunotherapy for the treatment of metastatic breast cancer patients, *Photochem. Photobiol. Sci.*, 10:817–821. DOI: 10.1039/c0pp00306a. 10

[44] Hahn, S. M., Fraker, D. L., Mick, R., Metz, J., Busch, T. M., Smith, D., Zhu, T., Rodriguez, C., Dimofte, A., Spitz, F., Putt, M., Rubin, S. C., Menon, C., Wang, H. W., Shin, D., Yodh, A., and Glatstein, E. 2006. A phase II trial of intraperitoneal photodynamic therapy for patients with peritoneal carcinomatosis and sarcomatosis, *Clin. Cancer Res.*, 12:2517–2525. DOI: 10.1158/1078-0432.ccr-05-1625. 10

[45] Banerjee, S. M., Malhorta, A., El-Sheikh, S., Tsukagoshi, D., Tran-Dang, M., Mosse, A., Parker, S., Davidson, T. I., Williams, N. R., Bown, S., and Keshtgar, M. R. 2017. Photodynamic therapy for the treatment of primary breast cancer: Preliminary results of a phase I/IIa clinical trial, *Cancer Res.*, 77. DOI: 10.1158/1538-7445.sabcs16-ot2-02-01. 10

[46] Christodoulides, N., Nikolidakis, D., Chondros, P., Becker, J., Schwarz, F., Rössler, R., and Sculean, A. 2008. Photodynamic therapy as an adjunct to non-surgical periodontal treatment: A randomized, controlled clinical trial, *J. Periodontol.*, 79:1638–1644. DOI: 10.1902/jop.2008.070652. 11

[47] Segarra-Vidal, M., Guerra-Ojeda, S., Valles, L. S., Lopez-Roldan, A., Mauricio, M. D., Aldasoro, M., Alpiste-Illueca, F., and Vila, J. M. 2017. Effects of photodynamic therapy in periodontal treatment: A randomized, controlled clinical trial, *J. Clin. Periodontol.*, 44:915–925. DOI: 10.1111/jcpe.12768. 11

[48] Tschen, E. H., Wong, D. S., Pariser, D. M., Dunlap, F. E., Houlihan, A., Ferdon, M. B., and Phase, IV ALA-PDT Actinic Keratosis Study Group. 2006. Photodynamic therapy using aminolaevulinic acid for patients with nonhyperkeratotic actinic keratoses of the face and scalp: Phase IV multicentre clinical trial with 12-month follow up, *Br. J. Dermatol.*, 155:1262–1269. DOI: 10.1111/j.1365-2133.2006.07520.x. 11

[49] Brown, D. M., Michels, M., Kaiser, P. K., Heier, J. S., Sy, J. P., Ianchulev, T., and AN-CHOR Study Group. 2009. Ranibizumab versus verteporfin photodynamic therapy for neovascular age-related macular degeneration: Two-year results of the ANCHOR study, *Ophthalmology*, 116:57–65. DOI: 10.1016/j.ophtha.2008.10.018. 11

[50] Gottesman, M. M. 2002. Mechanisms of cancer drug resistance, *Ann. Rev. Med.*, 53:615–627. DOI: 10.1146/annurev.med.53.082901.103929. 11

[51] Berg, K., Folini, M., Prasmickaite, L., Selbo, P. K., Bonsted, A., Engesaeter, B. Ø., Zaffaroni, N., Weyergang, A., Dietze, A., Maelandsmo, G. M., Wagner, E., Norum, O. J., and Høgset, A. 2007. Photochemical internalization: A new tool for drug delivery, *Curr. Pharm. Biotechnol.*, 8:362–372. DOI: 10.2174/138920107783018354. 11

[52] Wang, J. T., Giuntini, F., Eggleston, I. M., Bown, S. G., and MacRobert, A. J. 2012. Photochemical internalisation of a macromolecular protein toxin using a cell penetrating peptide-photosensitiser conjugate, *J. Control Release*, 157:305–313. DOI: 10.1016/j.jconrel.2011.08.025. 12

[53] Berg, K., Weyergang, A., Prasmickaite, L., Bonsted, A., Hogset, A., Strand, M. T., Wagner, E., and Selbo, P. K. 2010. Photochemical internalization (PCI): A technology for drug delivery, *Meth. Mol. Biol.*, 635:133–145. DOI: 10.1007/978-1-60761-697-9_10. 12

[54] Martinez de Pinillos Bayona, A., Moore, C. M., Loizidou, M., MacRobert, A. J., and Woodhams, J. H. 2015. Enhancing the efficacy of cytotoxic agents for cancer therapy using photochemical internalisation, *Int. J. Cancer*, 138:1049–1057. DOI: 10.1002/ijc.29510. 12, 13

[55] Weyergang, A., Berstad, M. E., Bull-Hansen, B., Olsen, C. E., Selbo, P. K., and Berg, K. 2015. Photochemical activation of drugs for the treatment of therapy-resistant cancers, *Photochem. Photobiol. Sci.*, 14:1465–1475. DOI: 10.1039/c5pp00029g. 12, 15

[56] Fu, A., Tang, R., Hardie, J., Farkas, M. E., and Rotello, V. M. 2014. Promises and pitfalls of intracellular delivery of proteins, *Bioconjug. Chem.*, 25:1602–1608. DOI: 10.1021/bc500320j. 12

[57] Mathews, M. S., Vo, V., Shih, E. C., Zamora, G., Sun, C. H., Madsen, S. J., and Hirschberg, H. 2012. Photochemical internalization-mediated delivery of chemotherapeutic agents in human breast tumor cell lines, *J. Environ. Pathol. Toxicol. Oncol.*, 31:49–59. DOI: 10.1615/jenvironpatholtoxicoloncol.v31.i1.60. 12

[58] Yaghini, E., Dondi, R., Tewari, K. M., Loizidou, M., Eggleston, I. M., and MacRobert, A. J. 2017. Endolysosomal targeting of a clinical chlorin photosensitiser for light-triggered delivery of nano-sized medicines, *Sci. Rep.*, 7:6059. DOI: 10.1038/s41598-017-06109-y. 12

[59] Meng, F., Cheng, R., Deng, C., and Zhong, Z. 2012. Intracellular drug release nanosystems, *Materials Today*, 15:436–442. DOI: 10.1016/s1369-7021(12)70195-5. 12

[60] Stahl, P. and Schwartz, A. L. 1986. Receptor-mediated endocytosis, *J. Clin. Invest.*, 77:657–662. DOI: 10.1172/jci112359. 12

[61] Salatin, S. and Yari Khosroushahi, A. 2017. Overviews on the cellular uptake mechanism of polysaccharide colloidal nanoparticles, *J. Cell Mol. Med.*, 21:1668–1686. DOI: 10.1111/jcmm.13110. 12

[62] Sultan, A. A., Jerjes, W., Berg, K., Hogset, A., Mosse, C. A., Hamoudi, R., Hamdoon, Z., Simeon, C., Carnell, D., Forster, M., and Hopper, C. 2016. Disulfonated tetraphenyl chlorin (TPCS2a)—induced photochemical internalisation of bleomycin in patients with solid malignancies: A phase 1, dose-escalation, first-in-man trial, *Lancet Oncol.*, 17:1217–1229. DOI: 10.1016/s1470-2045(16)30224-8. 14

[63] Olivecrona, H., 2019. Photochemical internalization: current clinical trials in cholangiocarcinoma, *17th International Photodynamic Association World Congress*, Cambridge, MA, *Proc. SPIE*, 11070–110703C. DOI: 10.1117/12.2528203. 14

[64] Hogset, A., Nedberge, A. G., Edwards, V., Hakerud, M., Olivecrona, H., and Otterhaug, T. 2019. Phase I clinical study for validation of photochemical internalisation (fimaVacc): A novel technology for enhancing cellular immune responses important for therapeutic effect of peptide- and protein-based vaccines, *17th International Photodynamic Association World Congress*, Cambridge, MA, *Proc. SPIE*, 11070–110703P. DOI: 10.1117/12.2526042. 14

[65] Haug, M., Brede, G., Hakerud, M., Nedberg, A. G., Gederaas, O. A., Flo, T. H., Edwards, V. T., Selbo, P. K., Hogset, A., and Halaas, O. 2018. Photochemical internalization of peptide antigens provides a novel strategy to realize therapeutic cancer vaccination, *Front Immunol.*, 9:650. DOI: 10.3389/fimmu.2018.00650. 14

[66] Weyergang, A., Selbo, P. K., Berstad, M. E., Bostad, M., and Berg, K. 2011. Photochemical internalization of tumor-targeted protein toxins, *Lasers Surg. Med.*, 43:721–733. DOI: 10.1002/lsm.21084. 15

[67] Norum, O. J., Gaustad, J. V., Angell-Petersen, E., Rofstad, E. K., Peng, Q., Giercksky, K. E., and Berg, K. 2009b. Photochemical internalization of bleomycin is supe-rior to photodynamic therapy due to the therapeutic effect in the tumorperiphery, *Photochem. Photobiol.*, 85:740–749. DOI: 10.1111/j.1751-1097.2008.00477.x. 15

[68] Lilletvedt, M., Tonnesen, H. H., Hogset, A., Sande, S. A., and Kristensen, S. 2011. Evaluation of physicochemical properties and aggregation of the photosensitizers TPCS2a

and TPPS2a in aqueous media, *Pharmazie*, 66:325–333. DOI: 10.1691/ph.2011.0337. 15

[69] Bull-Hansen, B., Berstad, M. B., Berg, K., Cao, Y., Skarpen, E., Fremstedal, A. S., Rosenblum, M. G., Peng, Q., and Weyergang, A. 2015. Photochemical activation of MH3-B1/rGel: A HER2-targeted treatment approach for ovarian cancer, *Oncotarget*, 6:12436–12451. DOI: 10.18632/oncotarget.3814. 15

[70] Selbo, P. K., Weyergang, A., Eng, M. S., Bostad, M., Mælandsmo, G. M., Høgset, A., and Berg, K. 2012. Strongly amphiphilic photosensitizers are not substrates of the cancer stem cell marker ABCG2 and provides specific and efficient light-triggered drug delivery of an EGFR-targeted cytotoxic drug, *J. Control Release*, 159:197–203. DOI: 10.1016/j.jconrel.2012.02.003. 15

[71] Olsen, C. E., Weyergang, A., Edwards, V. T., Berg, K., Brech, A., Weisheit, S., Høgset, A., and Selbo, P. K. 2017. Development of resistance to photodynamic therapy (PDT) in human breast cancer cells is photosensitizer-dependent: Possible mechanisms and approaches for overcoming PDT-resistance, *Biochem. Pharmacol.*, 144:63–77. DOI: 10.1016/j.bcp.2017.08.002. 15

[72] Spring, B. Q., Rizvi, I., Xu, N., and Hasan, T. 2015. The role of photodynamic therapy in overcoming cancer drug resistance, *Photochem. Photobiol. Sci.*, 201514(8):1476–1491. DOI: 10.1039/c4pp00495g. 15

[73] Lou, P. J., Lai, P. S., Shieh, M. J., Macrobert, A. J., Berg, K., and Bown, S. G. 2006. Reversal of doxorubicin resistance in breast cancer cells by photochemical internalization, *Int. J. Cancer*, 119:2692–2698. DOI: 10.1002/ijc.22098. 15

[74] Bostad, M., Berg, K., Høgset, A., Skarpen, E., Stenmark, H., and Selbo, P. K. 2013. Photochemical internalization (PCI) of immunotoxins targeting CD133 is specific and highly potent at femtomolar levels in cells with cancer stem cell properties, *J. Control Release*, 168:317–326. DOI: 10.1016/j.jconrel.2013.03.023. 15

CHAPTER 3

Photosensitizers and Therapeutic Agents Used in PDT and PCI

3.1 INTRODUCTION

For a photosensitizer to be suitable for use in PDT, it must exhibit chemical, photophysical as well as biological characteristics, that allow its uptake by a tumor. The photosensitizer must also demonstrate fast clearance and possess a large absorption peak at light wavelengths above 630 nm [1]. Although the use of dyes in the presence of light and oxygen for the killing of microorganisms and later treatment of superficial skin tumors begun in the 1900s, the photosensitizers mainly recognized for their use in PDT today, began being introduced in the 1970s [2]. The photosensitizers are divided into three classes known as first-generation, second-generation, and third-generation photosensitizers.

3.2 PHOTOSENSITIZERS USED IN PDT

3.2.1 FIRST-GENERATION PHOTOSENSITIZERS

Sensitizers classified as first-generation photosensitizers include haematoporphyrin derivatives (HpD) and Photofrin [3]. HpD tends to localize within tumors and result in a good tumoricidal response when activated by red light. As an established active component of HpD, Porfimer sodium (Photofrin), was the first photosensitizer to acquire approval for the treatment of recurring superficial papillary bladder cancers by the means of PDT [1] at 2 mg/kg concentration [4]. Photofrin has several absorption peaks, the weakest of which is found at wavelength of 630 nm [5]. Although light of 630 nm displays a weak absorption peak, it is still commonly used for the activation of photofrin in clinical settings as shorter light wavelengths impede deep tissue penetration [6]. This therefore limits the required light dose range for treatment of tumors to 100–200 J/cm² [7]. Despite being the most frequently used first-generation photosensitizer clinically, photofrin's potential has been undermined by its ability to cause long-term skin photosensitivity [8], which can last between 4–12 weeks [1].

3.2.2   SECOND-GENERATION PHOTOSENSITIZERS

In order to overcome the aforementioned limitations associated with first-generation photosensitizers, but still maintain the same degree of effectiveness in treating tumors as photofrin, numerous second-generation photosensitiers have been developed [1, 9]. In general, these photosensitizers have better chemical purity as well as capability to absorb light of longer wavelengths and cause significantly less post treatment skin photosensitization compared to photofrin [1]. Examples of second-generation photosensitizers include compounds such as benzoporphyrin derivative and temoporfin (mTHPC) which have a stronger ability to generate singlet oxygen [10].

Another effective alternative is the approved agent 5-aminolavulinic acid (ALA) which can be used for the treatment of cutaneous lesions in combination with red or blue light. Interestingly, this compound does not possess photosensitizing properties but exists as a natural haem precursor [11]. Haem is produced through the conversion of protoporphyrin IX (Pp IX), which is an efficient photosensitizer, into haem with the assistance of ferrochelatase [12]. Since the level of ferrochelatase activity is lower in tumors than normal tissues, administering ALA leads to a considerably increased level of PpIX in the tumor cells [12]. Furthermore, the absorption spectrum of PpIX is near to that of porfimer sodium and is therefore activatable by wavelengths close to 630 nm [1, 13].

The benefits of applying PDT with ALA instead of porfimer sodium include rapid PpIX clearance and thus reduced photosensitivity (normally 1–2 days) [1, 14], the availability of topical and oral application options which are useful for the treatment of skin cancer and oral cavity/digestive tract cancer, respectively, as well as better tumor selectivity [15, 16]. However, the highly hydrophilic nature of ALA restricts its penetration through cells [17]. This problem has been resolved via the creation of several ALA alkyl ester derivatives that are capable of entering the cells more efficiently [18].

In 2001, photosensitizer mTHPC (tempoporfin, Foscan) received approval for the treatment of head and neck cancers [19]. The potency of this photosensitizer is higher than that of ALA [1], or porfimer sodium [20] and only needs light dose of 20 J/cm^2 for tumor treatment [21]. Furthermore, in comparison to porfimer sodium and ALA, mTHPC absorbs a higher wavelength of light at 652 nm [22, 23] which enables its deeper penetration into the tissue [1].

Immense focus has gone toward the development of new photosensitizers that are more tumor specific, activatable by longer wavelengths of light and cause an overall shorter duration of photosensitivity [24, 25]. Among other photosensitizers that have been investigated in clinical studies are tin ethyl etiopurpurin (SnET2) [26], mono-L-aspartyl chlorin e6 (Npe6) [27], benzoporphyrin derivative (BPD) [28], and lutetium texaphyrin (Lu-Tex) [29]. The absorption peaks of these compounds lay at higher wavelengths of 660-nm [30], 664-nm [27], 690-nm [7], and 732-nm, respectively. Moreover, these sensitizers have shown to result in very mild and brief skin photosensitivity [1].

3.2.3   THIRD-GENERATION PHOTOSENSITIZERS

The study of third-generation photosensitizers has become very popular in recent years. These photosensitizers can be targeted to the cancer through antibody or nanoparticle conjugation in order to improve treatment selectivity [31–36]. The utilization of targeted nanoparticles for PDT has been an active area in particular.

3.3   PHOTOSENSITIZERS USED IN PCI

A suitable photosensitizer for PCI must be able to localize within the membranes of endosomes or lysosomes without translocating to the cytosol in the acidic conditions of the lysosomes [37]. The incorporation of carboxylic groups into the photosensitizers can aid this mechanism, however such groups also allow the penetration of the photosensitizer through the cell membrane. Therefore, the replacement of the carboxylic groups with sulfonate groups that maintain their negative charge at physiological pH allows the uptake of the photosensitizers through adsorptive endocytosis and prevents their penetration through cell membrane or protonation in the acidic environment of the lysosomes [38].

Various *in vitro* and *in vivo* PCI studies have employed $AlPCS_{2a}$ (aluminium phthalocyanine) (Figure 3.1a) as it has a stable azaporphyrin macrocycle that allows light absorption in the far red region (670 nm) and two adjacently substituted sulfonate groups that give the sensitizer amphiphilic properties [39–44].

Another sulfonated chlorin-based sensitizer, disulfonated tetraphenyl chlorin ($TPCS_{2a}$-Fimaporfin, Amphinex) (Figure 3.1c) was developed as an alternative to $AlPCS_{2a}$ due to its easy and reliable synthesis in large volumes and the presence of very few regioisomers. An additional advantage of $TPCS_{2a}$ is its enhanced photobleaching rate which results in reduced skin toxicity [45, 46]. All of these factors provide $TPCS_{2a}$ with the optimal photobiological and photophysical features required for PCI as shown in various studies [47–58]. The commercially available Meso-tetra-phenyl porphyrin disulfonate ($TPPS_{2a}$) (Figure 3.1b) is the porphyrin equivalent of $TPCS_{2a}$ and has been used in pre-clinical PCI studies [43, 59–65].

3.4   CHEMOTHERAPEUTIC DRUGS USED IN PCI

Macromolecular drugs such as siglec-3 (CD33) targeted immunoconjugate (Mylotarg), human epidermal growth factor receptor 2 (HER2) targeted antibody (Herceptin), epidermal growth factor receptor (EGFR) targeted antibody (Cetuximab), and interleukin-2 (IL-2) modified diphteria toxin fusion protein (Ontak) have been approved for clinical applications [62]. However, saporin, bleomycin, gelonin, and gemcitabine are the main chemotherapeutic agents used in PCI experimental studies. Out of these agents only bleomycin and more recently gemcitabine have reached the clinical trial stage for PCI investigation [42, 46, 59, 62, 66].

Saporin also called Saporin-S6 is a naturally existing, plant-derived toxin which acts as a Type 1 ribosome-inactivating protein (RIP). RIPs function through the removal of the A4324

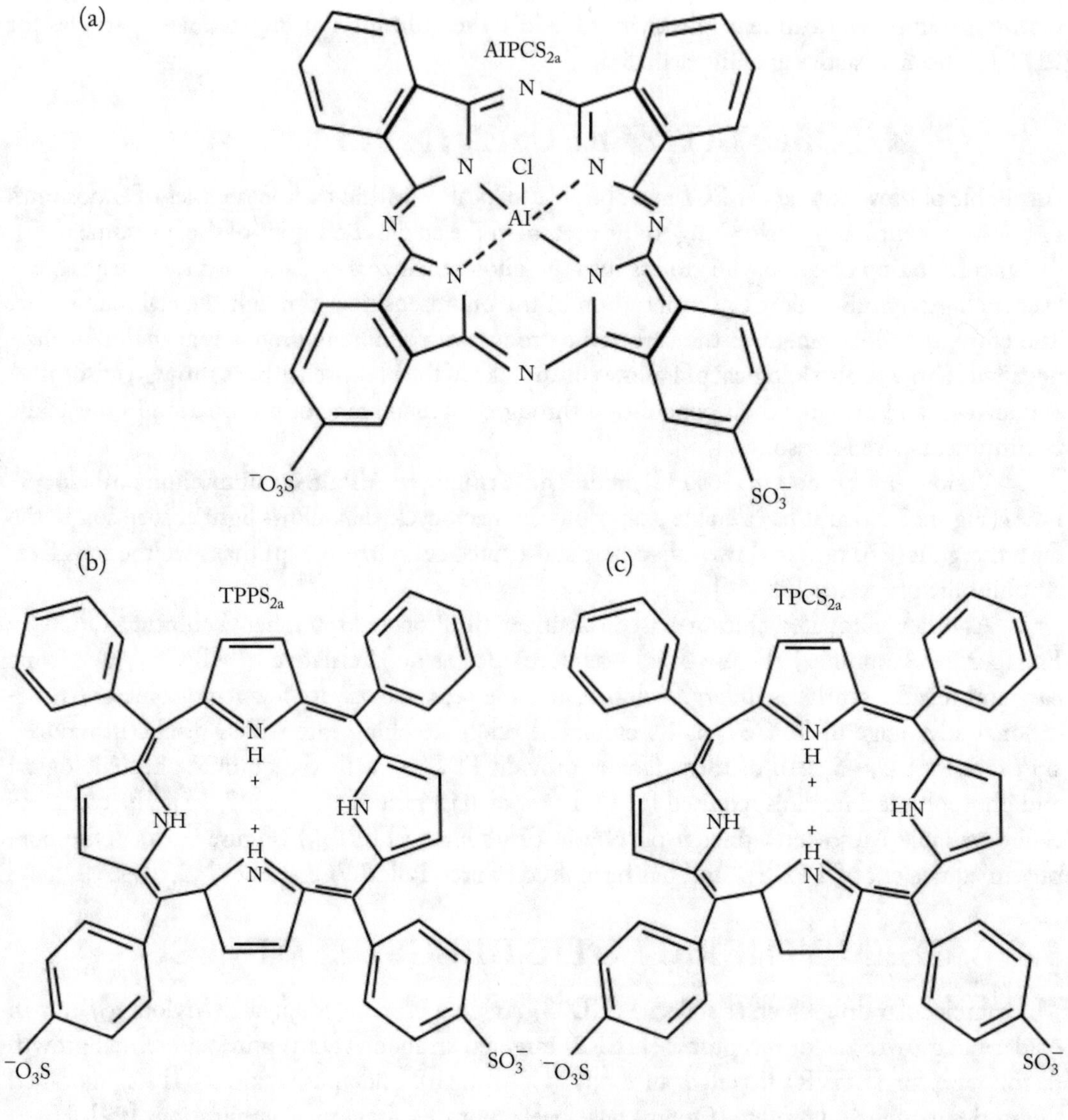

Figure 3.1: Structures of photosensitizers used in pre-clinical PCI studies. (a) AlPCS$_{2a}$, (b) TPPS$_{2a}$, and (c) TPCS$_{2a}$.

adenine residue, in the rat ribosome which subsequently interferes with the interaction between the ribosome and elongation factor 2. This causes irreversible damage to the ribosome and leads to inhibition of protein synthesis [67]. The high molecular weight of saporin (30 kDa) [68], makes it susceptible to endocytic uptake and endolysosomal degradation which limits its effectiveness when applied alone. Gelonin is also a plant derived RIP like saporin. The function of gelonin is therefore through removal of the base A4324 in 28S rRNA and preventing the association of elongation factor-1 and -2 with the 60S ribosomal subunit and, therefore, causing cell death. Like saporin, gelonin also displays limited cytotoxicity when applied alone since its high molecular weight (29 kDa) prevents it from crossing the plasma membrane and exerting its therapeutic effect [69, 70].

Bleomycin (Figure 3.2a), on the other hand, is a glycopeptide antibiotic which functions by producing a break in the double strand of DNA thus imitating the effects caused by ionizing radiation. The precise mechanism through which bleomycin functions is not fully clear, however; suggestions exist that this drug forms complexes with iron ions consequently forming a pseudoenzyme which could produce hydroxide and superoxide radicals that potentially cause breaks in DNA strands. Unfortunately, the therapeutic advantages of bleomycin can be limited by its toxic effects in the lung which result in fibrosis [71].

Dactinomycin (Figure 3.2b) is a clinically approved agent which acts as an anti-tumor antibiotic and functions through DNA intercalation as well as inhibition of RNA and protein synthesis. This agent has been assessed pre-clinically for PCI treatment of ovarian cancer and has proven to be efficacious at very low doses when used through this method. Due to its high molecular weight (approximately 1255 Da), dactinomycin uptake should occur partly via endocytosis like bleomycin which has a molecular weight of approximately 1400 Da. Furthermore, as a florescent compound, dactinomycin's endolysosomal localization can be detected via fluorescence imaging unlike bleomycin [72].

Doxorubicin has also been tested in pre-clinical PCI experiments in order to reverse doxorubicin resistance in breast cancer cells. The presence of an amino group in this drug contributes to its weak base properties and therefore makes it prone to protonation in the acidic lysosomes, thus leading to its entrapment as doxorubicin is less likely to penetrate through the lysosomal membrane and into the cytosol in an ionized form [73].

Gemcitabine is another anticancer drug that has been subjected to PCI delivery for the treatment of cholangiocarcinoma in a clinical study [66, 74]. Gemcitabine causes inhibition of ribonucleotide reductase which is a crucial enzyme involved in DNA synthesis. The inhibition of this enzyme prevents the cancer cells from copying their DNA and dividing properly and thus results in their death. Although it is a potent drug, Gemcitabine leads to myelosuppression as a common side effect [75, 76]. Therefore, delivering it at a low dose which causes cytotoxicity of the cancer cells while reducing or preventing severe side effects would have a clinical significance.

PCI has been used in several studies to enhance the delivery of targeted forms of Saporin and Gelonin [47–50, 63, 77, 78]. In one study, Weyergang et al. (2006) found that binding EGF

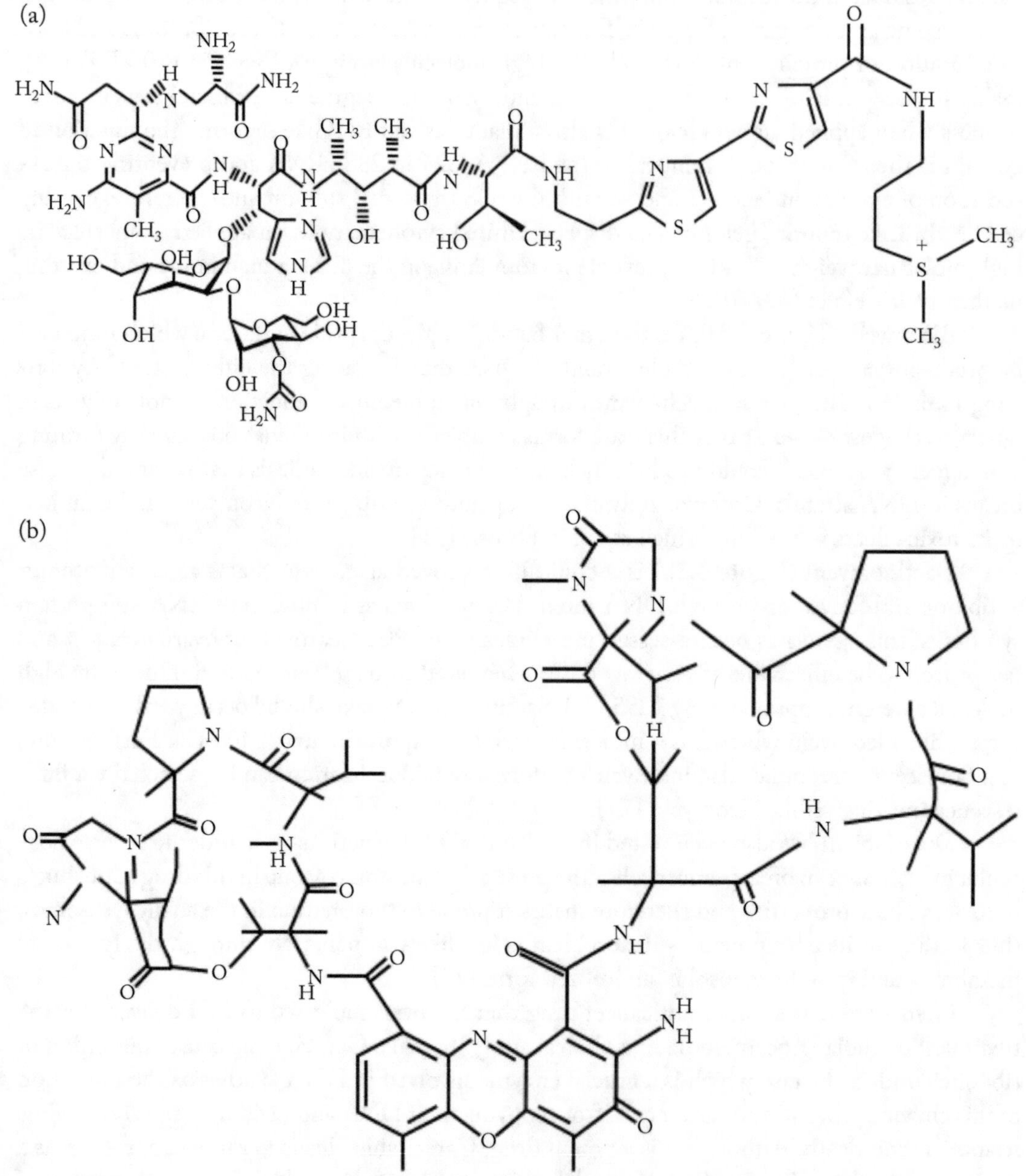

(a)

(b)

Figure 3.2: Examples of chemotherapeutic agents tested for use in PCI pre-clinically and clinically. (a) Structure of bleomycin which has been investigated in a PCI clinical trial. (b) Structure of Dactinomycin which has been assessed in a pre-clinical *in vitro* PCI study.

to saporin enhances PCI-induced cytotoxity by approximately 1000 folds in EGFR positive rat ovarian cancer cells (Nu Tu-19) than PCI alone which killed 50% of the cells [63]. The delivery of monoclonal Ig1 antibody (MOC31) bound gelonin via PCI was also shown to increase cytotoxicity in epithelial-glycoprotein-2 (EGP-2) expressing cancer cells compared to applying the immunotoxin alone [78].

In order to achieve the optimal size of protein macromolecules for efficient delivery and overcome the issues of renal excretion by smaller proteins as well as low diffusion rates into tumors and low penetration efficacy by larger proteins in *in vivo* models, recombinant targeted toxins were developed for using in combination with PCI. This process entails genetically fusing Type 1 RIP with a small-sized fragment of an antibody e.g., scFv or an endogenous ligand for the purpose of targeting [37]. Selbo et al. (2009) demonstrated the utilization of the first recombinant targeted toxin (scFvMEL/rGel) in combination with PCI for the targeting of the chondroitin sulfate proteoglycan 4 (CSPG4) *in vivo*. The results indicated that 50% of the mice that underwent PCI of scFvMEL/rGel had amelanotic melanoma tumors smaller than 800 mm^3 that did not increase in size for up to 110 days. Furthermore, 33% of the tumor-bearing mice showed complete regression after PCI of scFvMEL/rGel, compared to scFvMEL/rGel or PDT monotherapies which resulted in minor tumor growth delay thus implying that PCI of scFvMEL/rGel caused a synergistic effect [43].

Other recombinant targeted toxins used in PCI include vascular endothelial growth factor fused to recombinant gelonin (VEGF121/rGel) for targeting VEGF receptor 2 (VEGFR2), MH3-B1/rGel for targeting HER2, and EGF/rGel for targeting EGFR [51–54, 79]. PCI of VEGF121/rGel enhanced targeting of VEGFR2 in transfected endothelial cells (PAE/KDR) as well as improved treatment of murine colon carcinoma cells (CT26.CL25) *in vitro* and also led to induction of vascular collapse in colon cancer models *in vivo*. Furthermore, compared to PCI of bleomycin, VEGF121 PCI treatments leads to higher cure rates and was better tolerated *in vivo* [51]. PCI of MH3B1/rGel showed high efficacy against different breast cancer cell lines (MDA-MB-23, BT-20, Zr-75-1, SK-BR-3) with high and low HER2 expression levels [53], however such efficiency was not observed *in vivo* when compared to other targeted toxins [52]. Recombinant-targeted toxin EGF/rGel, on the other hand, was found to be highly efficient against EGFR expressing squamous carcinoma and melanoma cells *in vitro* as well as highly specific and potent against squamous tumors *in vivo* when delivered through PCI [54].

3.5   PCI FOR THE DELIVERY OF GENETIC MATERIAL INTO CELLS

PCI has been used for the efficient delivery of genetic material such as oligonucleotides, siRNA, mRNA, and plasmids which have on occasions been used for the purpose of silencing native genes. The large size and charge of nucleic acids prevents them from passing through plasma membrane of cells and their transport through delivery systems such as complexes and vectors are mostly impeded by the difficulty of escaping the endosome since most of these delivery systems are subject to endocytic uptake. Thus, PCI presents an extensive potential in such field [37].

As therapeutics, oligonucleotides are capable of down-regulating gene expression or modifying RNA splicing. Høgset et al. (2000) was the first group to demonstrate release of an oligonucleotide from endosomes using PCI. The study found that efficiency of transfection by DNA-poly-L-lysine complexes was enhanced by more than 20 fold in human melanoma cells with 50% of the surviving cells becoming transfected [80]. Other groups also followed to show that PCI-induced endosomal release results in increased biological activity of numerous oligonucleotides. Folini et al. (2003) utilized PCI for the effective release of naked peptide nucleic acid (PNA) targeted to the catalytic constituent of human telomerase reverse transcriptase (hTERT-PNA) into the cytoplasm of prostate cancer cells. The PCI delivery of hTERT-PNA led to a noticeable inhibition of telomerase activity as well as reduced cell survival compared to using hTERT-PNA alone [61].

The conjugation of PNAs to cationic peptides has also proven to improve cellular uptake and biological activity through PCI induced delivery [41, 81]. Shiraishi and Nielsen (2006) found that with some of the conjugates the biological activity was enhanced more 100 times in HeLa cells [41].

Nanocarriers and polymeric vehicles have also played an instrumental role in aiding the endosomal release of oligonucleotides, mRNA, siRNA, plasmids, and viral genes through PCI. These vehicles will be discussed further in Chapter 4.

3.6   VIRAL GENE DELIVERY THROUGH PCI

Viral systems have a good usability in the area of therapeutic gene delivery. Such delivery into cells allows transient expression and stable transgene integration into the genome of the target cell. Viral delivery vehicles have a higher efficiency than their non-viral counterparts mainly due to their enhanced cellular uptake and intrinsic endosomal escape functions as well as nuclear trafficking. However, due to concerns over safety of such systems, it would be advantageous to use low doses of the virus to execute virus-mediated gene delivery to the target sites. Using low doses of the virus would also help to reduce production issues and stimulation of unwanted immune reactions. Since most viruses are taken up by the cells via endocytosis and their mechanisms of releasing their genetic material from endosome are not always fully efficient, PCI-enhanced virus-mediated gene delivery has been investigated using two frequently used viral gene delivery systems (Adenovirus serotype 5 (Ad5) and Adenovirus associated virus (AAV)) which are prone to endocytosis [37].

After binding to coxackievirus and adenovirus receptor (CAR) receptor, Ad5 is normally endocytosed into the cell. Although it was believed that most viral particles are able to escape from the endosome after uptake, PCI surprisingly showed to increase adenovirus gene transduction by more than 20 folds in colon carcinoma cells compared to conventional adenovirus transduction. Furthermore, much lower doses of the viruses were required to cause effectiveness when PCI was employed [44, 82, 83].

As a commonly used vector AAV is also disposed to endosomal trafficking limitations due to its uptake via endocytosis [84, 85]. However, according to a study by Bonsted et al. (2005), PCI-enhanced AAV transduction in glioblastoma cells by 4 folds [65].

3.7 PCI IN CANCER VACCINATION

The development of an effective therapeutic cancer vaccination depends on the successful stimulation of cancer-specific CD8$^+$ cytotoxic T lymphocytes (CTLs) since these cells are known to be the most efficient at tumor cell killing. Prior to the activation of vaccine-induced CTL, antigens must be endocytosed into antigen presenting cells (APCs) for which the dendritic cells are the best candidates. The antigens are then processed in the cytosol by the major histocompatibility complex class I molecules (MHC I) presentation system and are eventually loaded onto MHC I. This leads to antigen cross-presentation to CD8$^+$ CTLs. Unfortunately, the endocytic uptake of peptide/protein antigens mainly results in their transportation into late endosomes and lysosomes which leads to the enzymatic degradation of the antigens. As a result, MHC II presentation and CD4$^+$ T lymphocyte activation occurs. While the CD4$^+$ T-cells are also crucial for anti-tumor immunity, results from pre-clinical studies have shown that CD8$^+$ CTLs stimulation is important for tumor cell elimination. Therefore, in order to avoid the dominance of the MHC class-II pathway, a system which allows antigens' endosomal escape to be enhanced is required so that MHC class-I cross-presentation as well as activation of effector and memory antigen-specific CD8$^+$ CTLs responses can be improved [86].

Various pre-clinical studies have explored the combination of PCI and anti-cancer vaccination for the purpose of delivering proteins and synthetic short or long peptide-based antigens into the cytosol [55–58].

Hakerud et al. (2014) co-injected protein antigen ovalbumin (OVA) and fimaporfin intradermally *in vivo*. After excitement of the photosensitizer using blue light and activation of PCI, CD8 T-cell response induction was found to be strongly increased and the growth of malignant murine melanoma cells were prevented compared to vaccination without PCI [55].

A subsequent study by the same group showed that PCI delivery of OVA causes induction, proliferation, as well as IFN-γ production of CTLs which in turn resulted in the suppression of melanoma tumor growth through the infiltration of the tumor by CTLs and caspase-3-dependent apoptosis [56].

3.8 REFERENCES

[1] Triesscheijn, M., Baas, P., Schellens, J. H., and Stewart, F. A. 2006. Photodynamic therapy in oncology, *Oncologist*, 11:1034–1044. DOI: 10.1634/theoncologist.11-9-1034. 23, 24

[2] Abrahamse, H. and Hamblin, M. R. 2016. New photosensitizers for photodynamic therapy, *Biochem. J.*, 473:347–364. DOI: 10.1042/bj20150942. 23

[3] Josefsen, L. B. and Boyle, R. W. 2008. Photodynamic therapy and the development of metal-based photosensitisers, *Metal-Based Drugs*, pages 1–23. DOI: 10.1155/2008/276109. 23

[4] Oseroff, A., Blumenson, L., Wilson, B., Mang, T., Bellnier, D., Parsons, J., Frawley, N., Cooper, M., Zeitouni, N., and Dougherty, T. 2006. A dose ranging study of photodynamic therapy with porfimer sodium (Photofrin®) for treatment of basal cell carcinoma, *Lasers Surg. Med.*, 38:417–426. DOI: 10.1002/lsm.20363. 23

[5] Usuda, J., Kato, H., Okunaka, T., Furukawa, K., Tsutsui, H., Yamada, K., Suga, Y., Honda, H., Nagatsuka, Y., Ohira, T., Tsuboi, M., and Hirano, T. 2006. Photodynamic therapy (PDT) for lung cancers, *J. Thorac. Oncol.*, 1:489–493. DOI: 10.1016/S1556-0864(15)31616-6. 23

[6] Vo-Dinh, T. 2015. *Biomedical Photonics Handbook*, 2nd ed., page 1–889, T. Vo-Dinh, CRC Press, Boca Raton, FL DOI: 10.1201/b17289. 23

[7] Allison, R. R., Downie, G. H., Cuenca, R., Hu, X. H., Childs, C. J., and Sibata, C. H. 2004. Photosensitizers in clinical PDT, *Photodiagn. Photodyn. Ther.*, 1:27–42. DOI: 10.1016/s1572-1000(04)00007-9. 23, 24

[8] Tanaka, M., Kataoka, H., Mabuchi, M., Sakuma, S., Takahashi, S., Tujii, R., Akashi, H., Ohi, H., Yano, S., Morita, A., and Joh, T. 2011. Anticancer effects of novel photodynamic therapy with glycoconjugated chlorin for gastric and colon cancer, *Anticancer Res.*, 31:763–769. 23

[9] O'Connor, A. E., Gallagher, W. M., and Byrne, A. T. 2009. Porphyrin and nonporphyrin photosensitizers in oncology: preclinical and clinical advances in photodynamic therapy, *Photochem. Photobiol.*, 85:1053–1074. DOI: 10.1111/j.1751-1097.2009.00585.x. 24

[10] Rajesh, S., Koshi, E., Philip, K., and Mohan, A. 2011. Antimicrobial photodynamic therapy: An overview, *J. Indian Soc. Periodontol.*, 15:323–327. DOI: 10.4103/0972-124x.92563. 24

[11] Yamashita, K., Hagiya, Y., Nakajima, M., Ishizuka, M., Tanaka, T., and Ogura, S. 2014. The effects of the heme precursor 5-aminolevulinic acid (ALA) on REV-ERBalpha activation, *FEBS Open Bio.*, 4:347–352. DOI: 10.1016/j.fob.2014.03.010. 24

[12] Wachowska, M., Muchowicz, A., Firczuk, M., Gabrysiak, M., Winiarska, M., Wańczyk, M., Bojarczuk, K., and Golab, J. 2011. Aminolevulinic acid (ALA) as a prodrug in photodynamic therapy of cancer, *Molecules*, 16:4140–4164. DOI: 10.3390/molecules16054140. 24

[13] Montcel, B., Mahieu-Williame, L., Armoiry, X., Meyronet, D., and Guyotat, J. 2013. Twopeaked 5-ALA-induced PpIX fluorescence emission spectrum distinguishes glioblastomas from low grade gliomas and infiltrative component of glioblastomas, *Biomed. Opt. Express.*, 4:548–558. DOI: 10.1364/boe.4.000548. 24

[14] Yano, T., Hatogai, K., Morimoto, H., Yoda, Y., and Kaneko, K. 2014. Photodynamic therapy for esophageal cancer, *Ann. Trans. Med.*, 2:29. DOI: 10.3978/j.issn.2305-5839.2014.03.01. 24

[15] Yang, D. F., Lee, J. W., Chen, H. M., and Hsu, Y. C. 2014. Topical methotrexate pretreatment enhances the therapeutic effect of topical 5-aminolevulinic acid-mediated photodynamic therapy on hamster buccal pouch precancers, *J. Formos. Med. Assoc.*, 113:591–599. DOI: 10.1016/j.jfma.2014.03.002. 24

[16] Ye, X., Yin, H., Lu, Y., Zhang, H., and Wang, H. 2016. Evaluation of hydrogel suppositories for delivery of 5-aminolevulinic acid and hematoporphyrin monomethyl ether to rectal tumors, *Molecules*, 21:1347. DOI: 10.3390/molecules21101347. 24

[17] Wan, M. T. and Lin, J. Y. 2014. Current evidence and applications of photodynamic therapy in dermatology, *Clin. Cosmet. Investig. Dermatol.*, 7:145–163. DOI: 10.2147/ccid.s35334. 24

[18] Rodriguez, L., de Bruijn, H. S., Di Venosa, G., Mamone, L., Robinson, D. J., Juarranz, A., Batlle, A., and Casas, A. 2009. Porphyrin synthesis from aminolevulinic acid esters in endothelial cells and its role in photodynamic therapy, *J. Photochem. Photobiol. B*, 96:249–254. DOI: 10.1016/j.jphotobiol.2009.07.001. 24

[19] De Visscher, S., Dijkstra, P., Tan, I., Roodenburg, J., and Witjes, M. 2013. mTHPC mediated photodynamic therapy (PDT) of squamous cell carcinoma in the head and neck: A systematic review, *Oral Oncol.*, 49:192–210. DOI: 10.1016/j.oraloncology.2012.09.011. 24

[20] Qumseya, B. J., David, W., and Wolfsen, H. C. 2013. Photodynamic therapy for Barrett's esophagus and esophageal carcinoma, *Clin. Endosc.*, 46:30–37. DOI: 10.5946/ce.2013.46.1.30. 24

[21] Hopper, C., Kubler, A., Lewis, H., Tan, I. B., and Putnam, G. 2004. mTHPC-mediated photodynamic therapy for early oral squamous cell carcinoma, *Int. J. Cancer*, 111:138–146. DOI: 10.1002/ijc.20209. 24

[22] Meier, D., Campanile, C., Botter, S. M., Born, W., and Fuchs, B. 2014. Cytotoxic efficacy of photodynamic therapy in osteosarcoma cells in vitro, *J. Vis. Exp.*, 51213. DOI: 10.3791/51213. 24

[23] Jerjes, W., Hamdoon, Z., and Hopper, C. 2012. Photodynamic therapy in the management of potentially malignant and malignant oral disorders, *Head Neck Oncol.*, 16. DOI: 10.1186/1758-3284-4-16. 24

[24] Olivo, M., Bhuvaneswari, R., Lucky, S. S., Dendukuri, N., and Soo-Ping Thong, P. 2010. Targeted therapy of cancer using photodynamic therapy in combination with multifaceted anti-tumor modalities, *Pharmaceuticals*, 3:1507–1529. DOI: 10.3390/ph3051507. 24

[25] Josefsen, L. B. and Boyle, R. W. 2008. Photodynamic therapy: Novel third-generation photosensitizers one step closer?, *Br. J. Pharmacol.*, 154:1–3. DOI: 10.1038/bjp.2008.98. 24

[26] Huang, Z. 2005. A review of progress in clinical photodynamic therapy, *Technol. Cancer Res. Treat.*, 4:283–293. DOI: 10.1177/153303460500400308. 24

[27] Yoon, I., Li, J. Z., and Shim, Y. K. 2013. Advance in photosensitizers and light delivery for photodynamic therapy, *Clin. Endosc.*, 46:7–23. DOI: 10.5946/ce.2013.46.1.7. 24

[28] Ormond, A. B. and Freeman, H. S. 2013. Dye sensitizers for photodynamic therapy, *Materials*, 6:817–840. DOI: 10.3390/ma6030817. 24

[29] Bhatta, A. K., Keyal, U., and Wang, X. L. 2016. Photodynamic therapy for onychomycosis: A systematic review, *Photodiagn. Photodyn. Ther.*, 15:228–235. DOI: 10.1016/j.pdpdt.2016.07.010. 24

[30] Poriel, C., Kessel, D., and Vicente, M. G. 2005. Stability of tin etiopurpurin, *Photochem. Photobiol.*, 81:149–153. DOI: 10.1111/j.1751-1097.2005.tb01534.x. 24

[31] Hudson, R., Carcenac, M., Smith, K., Madden, L., Clarke, O. J., Pelegrin, A., Greenman, J., and Boyle, R. W. 2005. The development and characterisation of porphyrin isothiocyanate-monoclonal antibody conjugates for photoimmunotherapy, *Br. J. Cancer*, 92:1442–1449. DOI: 10.1038/sj.bjc.6602517. 25

[32] Staneloudi, C., Smith, K. A., Hudson, R., Malatesti, N., Savoie, H., Boyle, R. W., and Greenman, J. 2007. Development and characterization of novel photosensitiser: scFv conjugates for use in photodynamic therapy of cancer, *Immunology*, 120:512-517. DOI: 10.1111/j.1365-2567.2006.02522.x. 25

[33] Maiolino, S., Moret, F., Conte, C., Fraix, A., Tirino, P., Ungaro, F., Sortino, S., Reddi, E., and Quaglia, F. 2015. Hyaluronan-decorated polymer nanoparticles targeting the CD44 receptor for the combined photo/chemo-therapy of cancer, *Nanoscale*, 7:5643–5653. DOI: 10.1039/c4nr06910b. 25

[34] Lamch, L., Bazylinska, U., Kulbacka, J., Pietkiewicz, J., Biezunska-Kusiak, K., and Wilk, K. A. 2014. Polymeric micelles for enhanced photofrin II (R) delivery, cytotoxicity and proapoptotic activity in human breast and ovarian cancer cells, *Photodiagn. Photodyn. Ther.*, 11:570–585. DOI: 10.1016/j.pdpdt.2014.10.005. 25

[35] Zhu, X., Wang, H., Zheng, L., Zhong, Z., Li, X., Zhao, J., Kou, J., Jiang, Y., Zheng, X., Liu, Z., Li, H., Cao, W., Tian, Y., Wang, Y., and Yang, L. 2015. Upconversion nanoparticle-mediated photodynamic therapy induces THP-1 macrophage apoptosis via ROS bursts and activation of the mitochondrial caspase pathway, *Int. J. Nanomed.*, 10:3719–3736. DOI: 10.2147/ijn.s82162. 25

[36] Kushibiki, T., Hirasawa, T., Okawa, S., and Ishihara, M. 2013. Responses of cancer cells induced by photodynamic therapy, *J. Healthc. Eng.*, 4:87–108. DOI: 10.1260/2040-2295.4.1.87. 25

[37] Jerjes, W., Theodossiou, T. A., Hirschberg, H., Hogset, A., Weyergang, A., Selbo, P. K., Hamdoon, Z., Hopper, C., and Berg, K. 2020. Photochemical internalization for intracellular drug delivery. From basic mechanisms to clinical research, *J. Clin. Med.*, 9:528. DOI: 10.3390/jcm9020528. 25, 29, 30

[38] Berg, K., Weyergang, A., Vikdal, M., Norum, O. J., Berstad, M., and Selbo, P. 2011. Photochemical internalization (PCI), a technology for site-specific drug delivery. Recent advances, *Photodiagn. Photodyn. Ther.*, 8:156. DOI: 10.1016/j.pdpdt.2011.03.111. 25

[39] Berg, K. and Moan, J. 1994. Lysosomes as photochemical targets, *Int. J. Cancer*, 59:814–822. DOI: 10.1002/ijc.2910590618. 25

[40] Prasmickaite, L., Hogset, A., Selbo, P. K., Engesæter, B. Ø., Hellum, M., and Berg, K. 2002. Photochemical disruption of endocytic vesicles before delivery of drugs: A new strategy for cancer therapy, *Br. J. Cancer* 86:652–657. DOI: 10.1038/sj.bjc.6600138. 25

[41] Shiraishi, T. and Nielsen, P. E. 2006. Photochemically enhanced cellular delivery of cell penetrating peptide-PNA conjugates, *FEBS Lett.*, 580:1451–1456. DOI: 10.1016/j.febslet.2006.01.077. 25, 30

[42] Berg, K., Dietze, A., Kaalhus, O., and Hogset, A. 2005. Site-specific drug delivery by photochemical internalization enhances the antitumor effect of bleomycin, *Clin. Cancer Res.*, 11:8476–8485. DOI: 10.1158/1078-0432.ccr-05-1245. 25

[43] Selbo, P. K., Rosenblum, M. G., Cheung, L. H., Zhang, W., and Berg, K. 2009. Multi-modality therapeutics with potent anti-tumor effects: Photochemical internalization enhances delivery of the fusion toxin scFvMEL/rGel, *PLoS One* 4:e6691. DOI: 10.1371/journal.pone.0006691. 25, 29

[44] Hogset, A., Engesaeter, B. O., Prasmickaite, L., Berg, K., Fodstad, O., and Mae-landsmo, G. M. 2002. Light-induced adenovirus gene transfer, an efficient and specific gene delivery technology for cancer gene therapy, *Cancer Gene Ther.*, 9:365–371. DOI: 10.1038/sj.cgt.7700447. 25, 31

[45] Berg, K., Nordstrand, S., Selbo, P. K., Tran, D. T., Angell-Petersen, E., and Hogset, A. 2011. Disulfonated tetraphenyl chlorin (TPCS2a), a novel photosensitizer developed for clinical utilization of photochemical internalization, *Photochem. Photobiol. Sci.*, 10:1637–1651. DOI: 10.1039/c1pp05128h. 25

[46] Sultan, A. A., Jerjes, W., Berg, K., Hogset, A., Mosse, C. A., Hamoudi, R., Hamdoon, Z., Simeon, C., Carnell, D., Forster, M., and Hopper, C. 2016. Disulfonated tetraphenyl chlorin (TPCS2a)—induced photochemical internalisation of bleomycin in patients with solid malignancies: A phase 1, dose-escalation, first-in-man trial, *Lancet Oncol.*, 17:1217–1229. DOI: 10.1016/s1470-2045(16)30224-8. 25

[47] Berstad, M. B., Weyergang, A., and Berg, K. 2012. Photochemical internalization (PCI) of HER2-targeted toxins: Synergy is dependent on the treatment sequence, *Biochim. Bio-phys. Acta.*, 1820:1849–1858. DOI: 10.1016/j.bbagen.2012.08.027. 25, 27

[48] Bostad, M., Kausberg, M., Weyergang, A., Olsen, C. E., Berg, K., Hogset, A., and Selbo, P. K. 2014. Light-triggered, efficient cytosolic release of IM7-saporin targeting the pu-tative cancer stem cell marker CD44 by photochemical internalization, *Mol. Pharm.*, 11:2764–2776. DOI: 10.1021/mp500129t. 25, 27

[49] Eng, M. S., Kaur, J., Prasmickaite, L., Engesaeter, B. O., Weyergang, A., Skarpen, E., Berg, K., Rosenblum, M. G., Maelandsmo, G. M., Hogset, A., Ferrone, S., and Selbo, P. K. 2018. Enhanced targeting of triple-negative breast carcinoma and malig-nant melanoma by photochemical internalization of CSPG4-targeting immunotoxins, *Photochem. Photobiol. Sci.*, 17:539–551. DOI: 10.1039/c7pp00358g. 25, 27

[50] Stratford, E. W., Bostad, M., Castro, R., Skarpen, E., Berg, K., Hogset, A., Myklebost, O., and Selbo, P. K. 2013. Photochemical internalization of CD133-targeting immuno-toxins efficiently depletes sarcoma cells with stem-like properties and reduces tumori-genicity, *Biochim. Biophys. Acta.* 1830:4235–4243. DOI: 10.1016/j.bbagen.2013.04.033. 25, 27

[51] Weyergang, A., Cheung, L. H., Rosenblum, M. G., Mohamedali, K. A., Peng, Q., Wal-tenberger, J., and Berg, K. 2014. Photochemical internalization augments tumor vascular cytotoxicity and specificity of VEGF(121)/rGel fusion toxin, *J. Control Release*, 180:1–9. DOI: 10.1016/j.jconrel.2014.02.003. 25, 29

[52] Bull-Hansen, B., Berstad, M. B., Berg, K., Cao, Y., Skarpen, E., Fremstedal, A. S., Rosenblum, M. G., Peng, Q., and Weyergang, A. 2015. Photochemical activation of MH3-B1/rGel: A HER2-targeted treatment approach for ovarian cancer, *Oncotarget*, 6:12436–12451. DOI: 10.18632/oncotarget.3814. 25, 29

[53] Bull-Hansen, B., Cao, Y., Berg, K., Skarpen, E., Rosenblum, M. G., and Weyergang, A. 2014. Photochemical activation of the recombinant HER2-targeted fusion toxin MH3-B1/rGel; Impact of HER2 expression on treatment outcome, *J. Control Release*, 182:58–66. DOI: 10.1016/j.jconrel.2014.03.014. 25, 29

[54] Berstad, M. B., Cheung, L. H., Berg, K., Peng, Q., Fremstedal, A. S., Patzke, S., Rosenblum, M. G., and Weyergang, A. 2015. Design of an EGFR-targeting toxin for photochemical delivery: In vitro and in vivo selectivity and efficacy, *Oncogene*, 34:5582–5592. DOI: 10.1038/onc.2015.15. 25, 29

[55] Hakerud, M., Waeckerle-Men, Y., Selbo, P. K., Kundig, T. M., Hogset, A., and Johansen, P. 2014. Intradermal photosensitisation facilitates stimulation of MHC class-I restricted CD8 T-cell responses of co-administered antigen, *J. Control Release*, 174:143–150. DOI: 10.1016/j.jconrel.2013.11.017. 25, 31

[56] Hakerud, M., Selbo, P. K., Waeckerle-Men, Y., Contassot, E., Dziunycz, P., Kundig, T. M., Hogset, A., and Johansen, P. 2015. Photosensitisation facilitates cross-priming of adjuvant-free protein vaccines and stimulation of tumour-suppressing CD8 T cells, *J. Control Release*, 198:10–17. DOI: 10.1016/j.jconrel.2014.11.032. 25, 31

[57] Varypataki, E. M., Hasler, F., Waeckerle-Men, Y., Vogel-Kindgen, S., Hogset, A., Kundig, T. M., Gander, B., Halin, C., and Johansen, P. 2019. Combined photosensitization and vaccination enable CD8 T-cell immunity and tumor suppression independent of CD4 T-cell help, *Front Immunol.*, 10:1548. DOI: 10.3389/fimmu.2019.01548. 25, 31

[58] Haug, M., Brede, G., Hakerud, M., Nedberg, A. G., Gederaas, O. A., Flo, T. H., Edwards, V. T., Selbo, P. K., Hogset, A., and Halaas, O. 2018. Photochemical internalization of peptide antigens provides a novel strategy to realize therapeutic cancer vaccination, *Front Immunol.*, 9:650. DOI: 10.3389/fimmu.2018.00650. 25, 31

[59] Martinez de Pinillos Bayona, A., Woodhams, J. H., Pye, H., Hamoudia, R. A., Moore, C. M., and MacRobert, A. J. 2017. Efficacy of photochemical internalisation using disulfonated chlorin and porphyrin photosensitisers: An in vitro study in 2D and 3D prostate cancer models, *Cancer Lett.*, 393:68–75. DOI: 10.1016/j.canlet.2017.02.018. 25

[60] Mohammad Hadi, L., Stamati, K., Loizidou, M., and MacRobert, A. J. 2018. Therapeutic enhancement of a cytotoxic agent using photochemical internalisation in 3D compressed collagen constructs of ovarian cancer, *Acta Biomaterialia*, 81:80–92. DOI: 10.1016/j.actbio.2018.09.041. 25

[61] Folini, M., Berg, K., Millo, E., Villa, R., Prasmickaite, L., Daidone, M. G., Benatti, U., and Zaffaroni, N. 2003. Photochemical internalization of a peptide nucleic acid targeting the catalytic subunit of human telomerase, *Cancer Res.*, 63:3490–3494. 25, 30

[62] Selbo, P. K., Weyergang, A., Bonsted, A., Bown, S. G., and Berg, K. 2006. Photochemical internalization of therapeutic macromolecular agents: A novel strategy to kill multidrug-resistant cancer cells, *J. Pharmacol. Exp. Ther.*, 319:604–612. DOI: 10.1124/jpet.106.109165. 25

[63] Weyergang, A., Selbo, P. K., and Berg, K. 2006. Photochemically stimulated drug delivery increases the cytotoxicity and specificity of EGF-saporin, *J. Control Release*, 111:165–173. DOI: 10.1016/j.jconrel.2005.12.002. 25, 27, 29

[64] Weyergang, A., Fremstedal, A. S., Skarpen, E., Peng, Q., Mohamedali, K. A., Eng, M. S., Cheung, L. H., Rosenblum, M. G., Waltenberger, J., and Berg, K. 2018. Light-enhanced VEGF121/rGel: A tumor targeted modality with vascular and immune-mediated efficacy, *J. Control Release*, 288:161–172. DOI: 10.1016/j.jconrel.2018.09.005. 25

[65] Bonsted, A., Hogset, A., Hoover, F., and Berg, K. 2005. Photochemical enhancement of gene delivery to glioblastoma cells is dependent on the vector applied, *Anticancer Res.*, 25:291–297. 25, 31

[66] Olivecrona, H. 2019. Photochemical internalization: Current clinical trials in cholangiocarcinoma, *17th International Photodynamic Association World Congress*, Cambridge, MA, *Proc. SPIE*, 11070–110703C. DOI: 10.1117/12.2528203. 25, 27

[67] Polito, L., Bortolotti, M., Mercatelli, D., Battelli, M. G., and Bolognesi, A. 2013. Saporin-S6ol.6: A useful tool in cancer therapy, *Toxins (Basel)*, 5:1698–1722. DOI: 10.3390/toxins5101698. 27

[68] Errico Provenzano, A., Posteri, R., Giansanti, F., Angelucci, F., Flavell, S. U., Flavell, D. J., Serena Fabbrini, M., Porro, D., Ippoliti, R., Ceriotti, A., Branduardi, P., and Vago, R. 2016. Optimization of construct design and fermentation strategy for the production of bioactive ATF-SAP, a saporin based anti-tumoral uPAR-targeted chimera, *Microb. Cell Fact.*, 15:194. DOI: 10.1186/s12934-016-0589-1. 27

[69] Rosenblum, M. G., Cheung, L. H., Liu, Y., and Marks, J. W. 2003. Design, expression, purification, and characterization, in vitro and in vivo, of an antimelanoma singlechain Fv antibody fused to the toxin gelonin, *Cancer Res.*, 63:3995–4002. 27

[70] Yuan, X., Lin, X., Manorek, G., and Howell, S. B. 2011. Challenges associated with the targeted delivery of gelonin to claudin-expressing cancer cells with the use of activatable

cell penetrating peptides to enhance potency, *BMC Cancer*, 11:61. DOI: 10.1186/1471-2407-11-61. 27

[71] Nicolay, N. H., Ruhle, A., Perez, R. L., Trinh, T., Sisombath, S., Weber, K. J., Ho, A. D., Debus, J., Saffrich, R., and Huber, P. E. 2016. Mesenchymal stem cells are sensitive to bleomycin treatment, *Sci. Rep.*, 6:26645. DOI: 10.1038/srep26645. 27

[72] Mohammad Hadi, L., Yaghini, E., MacRobert, A. J., and Loizidou, M. 2020. Synergy between photodynamic therapy and dactinomycin chemotherapy in 2D and 3D ovarian cancer cell cultures, *Int. J. Mol. Sci.*, 21:E3203. DOI: 10.3390/ijms21093203. 27

[73] Lou, P. J., Lai, P. S., Shieh, M. J., Macrobert, A. J., Berg, K., and Bown, S. G. 2006. Reversal of doxorubicin resistance in breast cancer cells by photochemical internalization, *Int. J. Cancer*, 119:2692–2698. DOI: 10.1002/ijc.22098. 27

[74] Martinez de Pinillos Bayona, A., Moore, C. M., Loizidou, M., MacRobert, A. J., and Woodhams, J. H. 2015. Enhancing the efficacy of cytotoxic agents for cancer therapy using photochemical internalisation, *Int. J. Cancer*, 138:1049–1057. DOI: 10.1002/ijc.29510. 27

[75] Toschi, L., Finocchiaro, G., Bartolini, S., Gioia, V., and Cappuzzo, F. 2005. Role of gemcitabine in cancer therapy, *Future Oncol.*, 1:7–17. DOI: 10.1517/14796694.1.1.7. 27

[76] Samanta, K., Setua, S., Kumari, S., Jaggi, M., Yallapu, M. M., and Chauhan, S. C. 2019. Gemcitabine combination nano therapies for pancreatic cancer, *Pharmaceutics*, 11:574. DOI: 10.3390/pharmaceutics11110574. 27

[77] Yip, W. L., Weyergang, A., Berg, K., Tonnesen, H. H., and Selbo, P. K. 2007. Targeted delivery and enhanced cytotoxicity of cetuximab-saporin by photochemical internalization in EGFR-positive cancer cells, *Mol. Pharm.*, 4:241–251. DOI: 10.1021/mp060105u. 27

[78] Selbo, P. K., Sivam, G., Fodstad, O., Sandvig, K., and Berg, K. 2000. Photochemical internalisation increases the cytotoxic effect of the immunotoxin MOC31-gelonin, *Int. J. Cancer*, 87:853–859. DOI: 10.1002/1097-0215(20000915)87:6%3C853::aid-ijc15%3E3.0.co;2-0. 27, 29

[79] Weyergang, A., Fremstedal, A. S., Skarpen, E., Peng, Q., Mohamedali, K. A., Eng, M. S., Cheung, L. H., Rosenblum, M. G., Waltenberger, J., and Berg, K. 2018. Light-enhanced VEGF$_{121}$/rGel: A tumor targeted modality with vascular and immunemediated efficacy, *J. Control Release*, 288:161–172. DOI: 10.1016/j.jconrel.2018.09.005. 29

[80] Hogset, A., Prasmickaite, L., Tjelle, T. E., and Berg, K. 2000. Photochemical transfection: A new technology for light-induced, site-directed gene delivery, *Hum. Gene Ther.*, 11:869–880. DOI: 10.1089/10430340050015482. 30

[81] Bøe, S. and Hovig, E. 2006. Photochemically induced gene silencing using PNA-peptide conjugates, *Oligonucleotides*, 16:145–157. DOI: 10.1089/oli.2006.16.145. 30

[82] Greber, U. F., Willetts, M., Webster, P., and Helenius, A. 1993. Stepwise dismantling of adenovirus 2 during entry into cells, *Cell*, 75:477–486. DOI: 10.1016/0092-8674(93)90382-z. 31

[83] Leopold, P. L., Ferris, B., Grinberg, I., Worgall, S., Hackett, N. R., and Crystal, R. G. 1998. Fluorescent virions: Dynamic tracking of the pathway of adenoviral gene transfer vectors in living cells, *Hum. Gene Ther.*, 9:367–378. DOI: 10.1089/hum.1998.9.3-367. 31

[84] Bantel-Schaal, U., Hub, B., and Kartenbeck, J. 2002. Endocytosis of adeno-associated virus type 5 leads to accumulation of virus particles in the Golgi compartment, *J. Virol.*, 76:2340–2349. DOI: 10.1128/jvi.76.5.2340-2349.2002. 31

[85] Hansen, J., Qing, K., and Srivastava, A. 2001. Adeno-associated virus type 2-mediated gene transfer: Altered endocytic processing enhances transduction efficiency in murine fibroblasts, *J. Virol.*, 75:4080–4090. DOI: 10.1128/jvi.75.9.4080-4090.2001. 31

[86] van der Burg, S. H., Arens, R., Ossendorp, F., van Hall, T., and Melief, C. J. 2016. Vaccines for established cancer: Overcoming the challenges posed by immune evasion, *Nat. Rev. Cancer*, 16:219–233. DOI: 10.1038/nrc.2016.16. 31

CHAPTER 4

The Use of Nanoparticles in PDT and PCI

4.1 INTRODUCTION

The employment of nanoparticles (NPs) in PDT and PCI has helped to overcome limitations such as poor biodistribution (e.g., due to poor water solubility) that are associated with utilization of current photosensitizers [1–3] and lack of applicability of PCI to small sized therapeutic molecules (~ 500 Dalton) such as doxorubicin, tamoxifen, etc. [4]. NPs are submicroscopic particles that are typically sized between 1–100-nm which are generally susceptible to endocytic uptake due to their sizes. NPs can be designed and produced out of various natural or synthetic materials and have the potential to carry multiple theranostic agents, through a targeted approach [5, 6]. It is possible to modify photosensitizers through incorporation into delivery systems such as liposomes, polymeric NPs, micelles, gold NPs, and ceramic NPs [7]. NPs employed for the purpose of improving PDT and PCI can be categorized according to their functional roles and whether they are actively or passively targeted [8].

Targeted NPs offer the advantage of improved selectivity as photosensitizers can be delivered to the tumor site while causing minimal harm to normal tissues. To "actively" target NPs to the cancer site, surface-conjugated ligands specific to overexpressed receptors or antigens on the target tissue can be used. The enhanced permeability and retention (EPR) effect also helps to improve the tumor-targeting properties of NPs in a process called "passive" targeting selectivity, where there is an increased selective uptake and retention of NPs with respect to normal adjacent tissue thus resolving issues such as poor bioavailability and unfavorable biodistribution that are associated with most clinically approved photosensitizers [9, 10]. The EPR is caused by irregular tumor neovasculature which has higher permeability than normal tissue microvasculature, as well as poorer tumor lymphatic drainage. The combination of both of these factors facilitates photosensitizer-carrier diffusion and retention within tumors [11]. In this case, photosensitizers can be targeted via encapsulation and or conjugation to nanocarriers [6].

Using nanocarriers in PDT could also offer several other advantages depending on the carrier type and mode of photosensitizer loading. Such advantages include the following. (1) Their large surface areas to volume ratios enable the amount of photosensitizer delivered to the target cells to be increased [12, 13]. (2) NPs could avert nonspecific photosensitizer accumulation in normal tissues by preventing the premature release of the sensitizer and its inactivation by plasma components. This could help to lessen the overall photosensitivity caused by the sensitizer [14]. (3) The increased amphilicity of the photosensitizers which results from the conjugation of the sensitizer to NPs allows the formulation to travel through the bloodstream unimpeded and localize within the tumor tissue [15]. (4) The systemic biodistribution, pharmacokinetics, cellular uptake, and targeting abilities of these carriers can be improved through further modification of their surfaces with functional groups or targeting agents [5, 16]. The actively and passively targeted NPs are subclassified based on their composition status as biodegradable or non-biodegradable.

Biodegradable NPs are composed of polymers that undergo enzymatic hydrolysis within biological settings therefore releasing the photosensitizers [10]. Nonbiodegradable NPs, e.g., silica do not necessarily require the liberation of photosensitizers from their bodies as free oxygen diffusion through the NPs allows singlet oxygen to be generated and released [2].

There are also non-biodegradable NP-based photosensitizers such as upconversion NPs, quantum dots, and self-lighting NPs which have inherent photosensitizing properties [7]. Examples of new NPs used for PDT and PCI are discussed in the next section.

4.2   ROLES OF DIFFERENT NANOPARTICLES IN PDT AND PCI

4.2.1   FULLERENES AND CARBON NANOTUBES (CNTS)

The structure of fullerenes (C_{60}) includes 60 carbon atoms that are arranged in a highly stable and spheroidal structure. Fullerenes have a diameter of approximately 2-nm and absorb wavelengths in the visible light region. These nanoparticles have high triplet yields as well as ROS-producing ability when illuminated [17]. One of the drawbacks associated with fullerenes is their poor solubility which may be overcome through approaches such as encapsulation within particular carriers, introduction of hydrophilic attachments or reduction of the molecules to water soluble anions, and suspension of fullerenes in the presence of co-solvents [18]. Another disadvantage of these NPs is that their highest absorption peak is located in the ultraviolet and blue regions which prevents them from penetrating deep into the tissue [19]. Thus, even though cancer treatment involving PDT and fullerenes has been shown *in vivo* [20–22], fullerenes with the capability to absorb red-near-infrared light must be developed, in order to be used effectively alongside PDT for clinical applications. Guan et al. (2016) developed *tri*-malonate derivative of fullerenes C_{70} (TFC_{70})/photosensitizer (Chlorin e6, Ce6) nanovesicles (FCNVs) that could absorb light in the near infrared region as a theranostic tool in mice breast tumor models. Maximum fluorescence intensity was detected from the FCNVs in the tumor site 4 hr after injection while the free Ce6 displayed negligible fluorescence indicating a significantly higher accumulation of FCNVs in the tumor than free Ce6. Following irradiation with 660-nm laser the FCNV-treated tumors turned black and scabby 7 days post treatment and became ablated 15 days after treatment while the tumors in the control groups continued to grow. Furthermore, no obvious toxicity to the mice was observed with the application of these nanovesicles and the FCNVs could undergo excretion by the liver and kidney with blood circulation over a long period [22].

Carbon nanotubes (CNTs) also consist of a framework of carbon atoms in the same way as fullerenes, however, they are rolled into a tube-like structure [23]. CNTs exist in two forms: (1) single-walled nanotubes (SWNTs) or (2) multi-walled nanotubes (MWNTs). The high surface area of these nanotubes allows drugs, peptide, and nucleic acid molecules to be incorporated into their walls as well as tips therefore enabling access to cancer cells either through endocytic crossing of the cell membrane or via recognition of cancer-specific receptors on the surface of cells [24]. Interestingly, SWNTs are also believed to be useful fluorescence probe quenchers [25]. Although the ability of SWNTs to improve the therapeutic efficacy of PDT has been demonstrated pre-clinically [26, 27], the limited knowledge available regarding the pharmacological and toxicological properties of CNTs, has prevented their utilization in a clinical setting to date.

4.2.2   QUANTUM DOTS (QDS)

Quantum dots (QDs) exist in form of nanocrystals which are a few nm in diameter and hold useful properties such as high fluorescence quantum yields and photostability [28]. These NPs possess greater absorption coefficients as well as size-tunable light emission and better signal brightness in comparison to most other fluorophores [29]. It is possible to develop water soluble and targeted QDs through applying specific modifications to the surface coating of the NPs. Cadmium selenide (CdSe) is the material frequently used to produce semiconductor QDs. In order to diminish the cytotoxicity caused by these NPs, cadmium-free QDs have been developed as an alternative where the cadmium is replaced by either nontoxic metals or metals with lower toxicities than cadmium, e.g., Indium (In) [30, 31]. The use of Indium-based QDs as a potentially more superior alternative to blue dyes for sentinel node mapping in breast cancer models has been demonstrated *in vivo* [32]. The photosensitizing properties of QDs can be further enhanced by conjugating them to photosensitizers (QDs-PS) [33]. QDs have been used in numerous PDT studies [34–37]. Ge et al. (2014) fabricated highly water dispensable graphene QDs (GQDs) that are activatable by light in the UV-visible region for PDT treatment of cervical and breast cancer *in vitro* and *in vivo*, respectively. For the *in vitro* studies, lasers of 405-nm and 637-nm wavelength were used while for the *in vivo* study light of 400–800-nm was used for the excitation of GQDs. The GQDs emitted strongly at 680-nm which also made them useful as imaging tools. Following application of GQDs and irradiation the cervical cancer cells showed nuclear condensation as well as morphological changes which included shrinkage and formation of blebs. Furthermore, in the presence of 0.036 μM GQDs, the cell viability was found to be 60% which then decreased to 20% by increasing the GQD concentration to 1.8 μM. However, such concentration of GQDs had little effect on cellular survival in the dark which was an indication of the low toxicity and biocompatibility of GQDs. In the *in vivo* experiments, the breast tumors began decomposing 9 days post treatment and eventually became destroyed

17 days after treatment. Moreover, the PDT treatment prevented tumor regrowth for up to 50 days and induced no obvious side effects [36].

4.2.3   UPCONVERSION NANOPARTICLES

Upconversion NPs (UCNPs) are a new generation of fluorophores, that are activatable by near infrared (NIR) radiation via a nonlinear optical mechanism [38]. The key material required for the production of UCNPs is $NaYF_4$ [39]. One form of UCNP which consists of $NaYF_4$ nanocrystal with mesoporous silica-loaded zinc phthalocyanine (ZnPc) coating has shown to have the ability to change NIR light into visible light upon experiencing excitement as result of exposure to NIR laser. Such property enables enhanced activation of the photosensitizer leading to release of $^1O_2$ and destruction of cancer cells. UCNPs have several advantages such as resistance to photobleaching and deeper tissue penetration as result of becoming excited by NIR light. Such benefits make UCNPs desirable for application in PDT as shown in various studies [40–43]. Furthermore, the utilization of self-lighting PDT which involves the employment of scintillation luminescent NP-attached photosensitizers has led to the improvement of PDT efficacy without the need for an external light source [44]. UCNPs have also proven suitable for PCI applications involving drug and oligonucleotide (for gene silencing) delivery [45–47]. Wang et al. (2011) treated breast tumors in mice intratumorally with 40–50 μL of UCNP-Ce6 (20 mg/mL UCNP, ~1.5 mg/mL Ce6) and NIR laser (980-nm). Out of all the tumor models, 70% disappeared 2 weeks post treatment and exhibited a survival of over 60 days as well as no re-growth, whereas applications of UCNP-Ce6 alone and NIR laser alone were unable to delay tumor growth. The remaining 30% of the treated tumor models became partially damaged and demonstrated far slower growth rates than the control groups as well as a survival of 32–42 days. Impressively, the use of NIR laser considerably increased tissue penetration depth in the models. Furthermore, the UCNPs were found to clear out of the mouse organs slowly following PDT treatment without causing substantial toxicity to the animals [40]. Jayakumar et al. (2014) also used NIR light (980-nm) along with core-shell UCNPs in their PCI study to knockdown STAT3 in mouse skin melanoma. The UCNPs in this study were able to convert the 980-nm light into blue light to excite $TPPS_{2a}$ (413-nm) for the purpose of PCI as well as into UV light that enabled cleavage of photolabile tether thus leading to the release of STAT3 antisense morpholino and eventually the knockdown of STAT3. The study showed that light activation of UCNPs resulted in an increase in STAT3 knockdown both in the presence and absence of $TPPS_{2a}$ (PCI), however such elevation was higher in the presence of PCI. Furthermore, the UCNP treatment led to significant growth inhibition particularly in the presence of PCI [47].

4.3   BIODEGRADABLE NANOPARTICLES

Biodegradable polymer-based NPs have demonstrated a vast potential as photosensitizer carriers due to their ability to control drug release, their versatile properties in material manufacturing processes as well as their great drug loading capabilities. The chemical composition and architecture of the polymers can be adjusted to suit, various photosensitizers with different hydrophobicity degrees, molecular weights, pH, and charges [48]. Polymer degradation can be induced by the lower tumor pH or enzymatically depending on the expression level of the particular enzyme by the tumor. The NPs may also be actively targeted to the specific sites of action through surface modifications. Photosensitizers can be delivered via micelles, liposomes, dendrimers, as well as NPs [49]. In general, NPs are composed of natural or synthetic polymers. Examples of naturally occurring polymers are biopolyesters such as poly(hydroxyalkanoates) (PHAs), poly(β-amino esters) (PbAE), and poly(α-hydroxy esters); while synthetic polymers include poly(lactide) (PLA), poly(glycolide) (PGA), poly(glycolide-co-lactide) and poly(ε-caprolactone) (PCL), all of which are biocompatible and used extensively for the preparation of NPs and micelles [5]. Pre-clinical studies using polymeric micelles and NPs for photosensitizer delivery in PDT showed that the targeting ability and pharmacokinetic functions of these delivery systems are controlled by their composition, size, surface charge, morphology, and hydrophobicity. A majority of second-generation photosensitizers have been tested with the FDA-approved PLGA [50–57]. An example of photosensitizer encapsulation in liposome has been demonstrated in Figure 4.1b.

4.3.1   MICELLES AND PLGA NANOPARTICLES

Folate conjugated m-THPC-loaded micelles were developed by Syu et al. (2012) to improve tumor targeting of m-THPC (photosensitizer) and increase treatment efficacy while sparing the healthy tissue from photodamage. The conjugated micelles were found to be uptaken by cervical cancer cells over-expressing the folate receptor both *in vitro* and *in vivo* with the PDT having no substantial unfavorable effects on the body weight of the mice. Extending the drug delivery period *in vivo* resulted in greater tumor growth inhibition (92%) after irradiation compared to the free m-THPC or folate free m-THPC loaded micelles. Folate conjugation was also found to reduce the amount of m-THPC required for PDT [50].

Loureiro da Silva et al. (2013), on the other hand, tested the topical application of Protoporphyrin IX (PpIX) loaded PLGA NPs on mice skin. Through fluorescence microscopy it was determined that the florescence intensity detected 24 hr post application in deeper regions of the tissue was greater than 8 hr post application. This indicated that the sensitizer-NP conjugate localized in the epidermis and dermis of the skin which is known as a site of action for topically applied PDT [57].

These studies have demonstrated different degrees of improvement in photodynamic activity, tumor growth inhibition, ROS production, as well as plasma circulation time which could allow photosensitizer administration levels to be decreased thus reducing side effects associated with the photosensitizer.

Micelles and PLGA NPs have also been employed in various PCI studies [58–62]. Tian et al. (2017) synthesized a pH sensitive amphiphilic co-polymer block (PEG$_{113}$-b-PCL$_{54}$-a-porphyrin) which self-assembled into micelles and functioned both as a photosensitizer and a doxorubicin carrier for PCI delivery to pulmonary cancer cells. Using light of 420-nm resulted in a slight PCI effect of doxorubicin at drug concentration of above 1.56 $\mu g/mL$ while no significant PDT effect was observed at sensitizer concentrations of up to 21.1 $\mu g/mL$. Such low degree PDT effect has been attributed to the increased sequestration of the polymers within the matrix of the endocytic vesicle. However, the singlet oxygen produced potentially by porphyrins in the micelles upon irradiation may cause disruption of the lysosomes thus helping doxorubicin to escape and enhance the therapeutic effect of the treatment [58]. Enhancement in singlet oxygen generation via utilization of Ce6 (photosensitizer) and doxorubicin loaded amphiphilic micelles was also observed by Park et al. (2014) when the formulation was compared to the application of free Ce6. According to the results, PCI using low dose of 670-nm laser resulted in overcoming of drug resistance in colon carcinoma cells *in vitro* and *in vivo* through lipid peroxidation of cellular membrane which considerably increases the uptake of doxorubicin without causing any undesirable side effects [59].

4.3.2   LIPOSOMES

Liposomal carriers have also gained a lot of interest over the years owing to their triggered release mechanism which enables them to liberate photosensitizers upon exposure to multimodal internal (i.e., pH or enzyme) or external (light or temperature) stimuli [5, 63–66]. Such factors could improve photosensitizer internalization within cells and subsequently contribute toward enhanced tumor accumulation thus minimizing non-specific photosensitivity of normal tissues [5]. In a study by Reshtov et al. (2013), two liposomal formulations (Foslip and Fospeg) were compared in terms of their pharmacokinetics, drug release, liposome stability, tumor uptake, intratumoral distribution, and impact on the effectiveness of temoporfin-PDT at different light—drug intervals in colorectal cancer *in vivo* models. The pegylated liposomal formulation "Fospeg" was found to be much greater than Foslip in terms of causing enhanced permeability and retention-based tumor accumulation, maintaining stability within blood circulation as well as enhanced photosensitizer release properties that resulted in better treatment efficacy. Furthermore, Fospeg required a significantly reduced light-drug interval to cause such high treatment efficacy. Foslip, however, was rapidly removed from the circulation because of premature drug release as well as destruction of the liposome [67].

Multifunctional liposomal formulations such as porphysomes have also been focus of studies due to their usefulness for theranostic applications. Porphysomes are composed of photosensitizer-phospholipid conjugates that are self-assembled into liposome like structures. This liposomal formulation has potentials as drug delivery, imaging, and therapeutic tools. These formulations absorb light in NIR region and have photothermal in addition to photoacoustic properties which allow cancer cells circulating in blood vessels and residing in sentinel lymph nodes to be detected in a non-invasive manner while also maintaining their drug delivery capabilities. The destabilization of porphysomes following endosomal uptake is manifested through enhanced fluorescence levels which indicate that these formulations have reached their target site [68, 69]. According to Lovell et al. (2011) porphysomes can be loaded actively or passively. This group also showed that using porphysomes allows the lymphatic system to be visualized in detail using photoacoustic tomorgraphy. The systemic administration of porphysomes led to their accumulation in cervical tumors in mice which then resulted in induction of tumor abla-tion, following irradiation with 658-nm laser 24 hr later. Furthermore, the biodegradable nature of the porphysomes prevented them from causing significant acute toxicity in the mice [68].

Liposomes have demonstrated pronounced competency as carriers in PCI studies as well [70–72]. Yaghini et al. (2018) employed a cell penetrating peptide (CPP)-modified lipo-some which was conjugated to photosenstiser meso-tetrakis tetraphenylporphyrin (TPP) and loaded with cytotoxin (Saporin) to enhance treatment of fibrosarcoma cells *in vitro*. Such PCI delivery of saporin improved cytotoxity levels in the fibrosarcoma cells with the effect increas-ing with rise in saporin concentration and irradiation dose [70]. In another study, phototoxin Hypericin (HYP) was encapsulated in liposomes and investigated for the treatment of prostate cancer cells. Encapsulating HYP in liposomes strongly enhanced the uptake of HYP by the cancer cells. The presence of the cationic charges from the guanidinium bearing lipids on the surface of liposome made them more prone to adsorptive endocytosis thus improving their cel-lular uptake and phototoxicity. However, the distance of the guanidinium group charges from the surface of the liposome greatly influenced HYP loading, subcellular localization, and pho-totoxicity [71].

4.3.3   DENDRIMERS

Another polymer-based NP that has attracted attention as a drug delivery tool with the poten-tial to enhance PDT and PCI is dendrimer. Dendrimers are known as three-dimensional (3D), monodispersed macromolecules that are size tuneable and have functional peripheral groups as well as an inner cavity which enable a range of molecules to be incorporated [5]. Their prepara-tion process entails a stepwise procedure which results in extremely ordered branching patterns which possess a consistent structure. The 3D structure and comparatively large size of den-drimers allows them to become internalized in organelles with limited membranes thus realiz-ing controlled localization intracellularly [73, 74]. The surface groups present on the dendrimers allow targeting molecules and functional groups to be attached to them. Furthermore, it is pos-sible to regulate the size and lipophilicity of dendrimer conjugates for the purpose of optimizing cellular uptake as well as tissue biodistribution [75, 76]. The quantity of branching available in a dendrimer is defined as generation number. In general, the central core molecule which has sites that can expand into additional branches is called generation 0 (G0). The generation number of each new branched point increases by 1 in an orderly manner (e.g., G1, G2, G3) [74].

All of these properties make dendrimers drug-carriers of interest for PDT studies. Pho-tosensitizers can either be covalently linked to functional groups present on the surface of the dendrimer or be encapsulated within the core of the dendrimer. Sensitizers that are tagged onto the exterior of the dendrimers are mostly released either post photoexcitation and activation of

ROS which dissolve the covalent bond existing between the sensitizer and dendrimer or via cellular esterases which hydrolyse these bonds [5]. Conjugation of the photosensitizer to the dendrimer surface not only reduces toxicity of this nanocarrier but also allows loading of the photosensitizer in monomeric form, precludes the untimely release of the photosensitizer and elevates photosensitizer accumulation intracellularly compared to free sensitizer [77–80].

Polyamidoamide (PAMAM) has been the most investigated dendrimer in PDT and has often been used for oligonucleotide and plasmid delivery in PCI [81–86]. One of the reasons for the frequent utilization of PAMAM in such studies is the commercial availability of PAMAM G0-G10 which possess a variety of peripheral and end groups as well as different molecular weights [74].

Narsireddy et al. (2015) used PAMAM (G4) which was conjugated to either near in-frared light sensitive photosensitizer or photosensitizer as well as a human epidermal growth factor receptor 2 (HER2) targeting peptide ligand (Affibody) to treat human HER2 overex-pressing ovarian and HER2-negative breast cancer *in vitro* through PDT. *In vivo* models of the HER2 overexpressing ovarian cancer were also treated with PDT using the two conjugated formulations. The results found that photosensitizers conjugated to dendrimers were more ef-ficient at causing PDT mediated cytotoxicity in HER2 positive ovarian cancer cells than free photosensitizers. However, such result was not observed in the HER2 negative breast cancer cells. In the *in vivo* models, the photosensitizer-dendrimer formulation successfully resulted in significant tumor suppression compared to free sensitizer. The addition of Affibody was found to further enhance the photodynamic therapeutic effects of the sensitizer, however, in the *in vivo* study, the improvement made by the Affibody conjugated formulation was marginal when compared to its Affibody lacking counterpart [83].

Yuan et al. (2015), on the other hand, tested numerous PAMAM dendrimers complexed with sensitizer Ce6 as well as phosphorodiamidate morpholino oligomer (a therapeutic third generation splice switching oligonucleotide (SSO)) for the simultaneous delivery of SSOs and photosensitizers into endolysosomal compartments and treatment of melanoma cells with the aid of PCI. According to the results, irradiation of the cells led to PCI mediated cytotoxicity. While the complex did not demonstrate any obvious dark cytotoxicity in the absence of illu-mination, an almost 20% reduction in cell viability was observed along with the effective PCI mediated SSO cytosolic delivery upon photo-irradiation. The involvement of PCI enabled de-struction of endolysosomal membranes by ROS thus leading to the release of splice switching oligonucleotides [85].

Another study by Shieh et al. (2008) used dendrimer, PAMAM (G4) as a gene carrier for cellular transfection. In this study, a photosensitizer activatable by blue light was conjugated to PAMAM which was then complexed with plasmid DNA and applied to cervical cancer cells *in vitro*. The sensitizer-dendrimer conjugate was found to be taken up by the ovarian cancer cells through endocytosis and accumulate in the endolysosomal compartments. Enhanced transfection of cells (26.7%) was achieved at non-toxic conditions upon irradiation compared to no irradiation (less than 3.5%) [86].

4.3.4   NATURAL POLYMER-BASED NANOPARTICLES

NPs consisting of natural cross-linked polymers have also been considered and investigated as photosensitizer carriers in PDT studies [87–89]. Examples of natural polymer NPs include proteins and polysaccharides.

4.3.5   HUMAN SERUM ALBUMIN NANOPARTICLES

Human serum albumin (HSA) has many potentials as a nanocarrier due to its good biocompatibility as well as non-toxic, and non-immunogenic properties. HSA NPs have high binding capacities for various drugs and can be simply synthesized through various methods [90, 91]. Furthermore, HSA-based NPs provide a means for delivering hydrophobic drugs without the requirement for potential toxic solvents. Given that this technology makes use of endogenous albumin pathways to selectively transport large quantities of chemotherapeutic agents to tumor sites and has great usability for the delivery of different agents particularly lipophilic drugs, it is no surprise that attention has been drawn toward HSA NPs as potential theranostic drug carriers [92]. Abraxane (nab-paclitaxel) was the first albumin NP-bound chemotherapeutic drug to receive approval from FDA for treating metastatic breast cancer [93]. Albumin based nanocarriers have been used by Annegret et al. (2011) for the delivery of lipophilic photosensitizers mTHPP and mTHPC. The NPs were formed from HSA prior to the loading of the photosensitizers onto them through an absorptive drug loading method. mTHPC-HSA NPs showed the highest intracellular uptake by lymphocytic cells after 1 hr of incubation, however, while mTHPC-HSA NPs showed a nearly similar intracellular uptake as mTHPC at 5 hr, the uptake of mTHPP-HSA NPs was lower than that of mTHPP, mTHPC, and mTHPC-HSA NPs. Both NPs generate singlet oxygen at 5 hr, however, the amount produced by mTHPC-HSA NPs was higher than that produced by mTHPP-HSA NPs. Interestingly, the rate of apoptosis caused by mTHPP-HSA NPs after 1 hr incubation followed by irradiation is considerably higher than that induced by mTHPC-HSA NPs, but it is lowered with increasing incubation times. However, mTHPC-HSA NPs led to a higher phototoxic effect (70–85%) than other treatment conditions between 5–24 hr incubation. The greater phototoxicity caused by mTHPC-HSA NPs compared to their mTHPP counterparts after 5 and 24 hr incubation is consistent with their higher intracellular uptake which was measured at those time points. Overall, an efficient singlet oxygen production as well as an effective cell death was observed upon incubation of cells with both mTHPP and mTHPC-HSA NPs [87]. There are some issues related to albumin nanocarrier development such as complex manufacturing methods, extensive size distribution, instability in physiological conditions, as well as early and unintentional photosensitizer release, which reduces its accumulation in the target site [94]. An alternative conjugation method was devised by Jeong et al. (2011) to bind Ce6 to lysine residues within

HSA in order for them to form self-assembled NPs. Upon irradiation, the Ce6 conjugated HSA NPs enhanced tumor-specific biodistribution and tumor reduction (70%) compared to free Ce6 (34%) in mice colorectal cancer models. The higher tumor specificity was found to be largely due to prolonged circulation of Ce6-HSA NPs in the blood as well as EPR effect [88].

4.3.6 CHITOSAN

The polysaccharide, chitosan, is produced through partial deacetylation of chitin which is obtained mainly from the shells of crustacean and insects. Chitosan becomes soluble and positively charged under acidic conditions and has been used for the development of amphiphilic-chitosan based nanocarriers for the delivery of hydrophobic anticancer drugs [95–99] and photosensitizers for therapy and imaging [100–103]. Photosensitizer loading onto chitosan NPs has either been carried out through encapsulation within the core of the self-assembled chitosan NP or conjugated to chitosan which then self-assembles into a NP [104–107]. The linking of photosensitizer to chitosan as described in the second method allows "on/off" carrier systems to be developed. Lee et al. (2011) demonstrated that, in the off state, the PpIX-chitosan NPs show no fluorescence or phototoxicity upon light exposure prior to cellular uptake. However, once the conjugates were taken up by squamous carcinoma cells and the PpIX was released, strong fluorescence signal as well as singlet oxygen were generated following irradiation. Moreover, PpIX-chitosan NPs showed improved tumor-targeting ability in addition to therapeutic efficiency compared to free PpIX in colorectal cancer mice models [106]. Loading of Ce6 onto chitosan NPs also enhanced PDT efficiency in alveolar adenocarcinoma cells when compared to free Ce6. Such loading also helped to improve the biocompatibility of Ce6 that is otherwise known to be toxic [101].

In a PCI study by Gaware et al. (2017), the conjugation of *meso*-Tetraphenylchlorin to chitosan, led to considerable increase in plasmid DNA transfection in human colon carcinoma cells upon activation by red light. The nanoconjugates also showed to induce a strong PDT as well as PCI effect (with bleomycin) *in vivo* [103].

4.3.7 LIPOPROTEIN AND DEXTRAN NANOPARTICLES

Low-density lipoprotein and dextran NPs have also been evaluated in pre-clinical PDT and PCI studies. The overexpression of LDL receptors in many malignant tumors has raised interest in LDL NPs as means of enhancing tumor cell delivery. Marotta et al. (2011) developed reconstituted bacteriochlorin e6 bisoleate low-density lipoprotein (r-Bchl-BOA-LDL) NPs in order to improve PDT efficacy in human hepatoblastoma G2 (HepG2) tumors *in vivo*. The ability of Bacteriochlorophyll (a magnesium chelate) to undergo activation by light in near-infrared region (748-nm) makes it desirable for use as a photosensitizer. The study revealed that r-Bchl-BOA-LDL-NPs caused a significant delay in tumor regrowth when injected at doses of 2 μmole/kg and illuminated with light fluences of 125, 150, or 175 J/cm^2 [108]. Jin et al. (2011), on the other hand, loaded fluorescent dyes on LDL NPs through three different approaches of

(1) surface loading, which involved intercalation of the dye into the phospholipid monolayer exterior, (2) protein loading, which involved conjugation of the dye to amino acids of apo-B-100 protein, or (3) core loading, which entailed reconstitution of the dye into the hydrophobic core of LDL. The incubation of cells with these cargo-loaded LDL NPs (CLLNPs) and photosensitizer (AlPcS$_{2a}$) followed by laser irradiation resulted in the efficient cytosolic discharge of surface-loaded and protein-labeled cargo [109].

Dextran is a polysaccharide polymer which has chain units of different lengths. As a biocompatible material, dextran has commonly been employed in biomedical applications for the coating of NPs, in particular magnetic NPs (iron oxide) that have also been used in pre-clinical PDT studies for magnetic resonance imaging and treatment of cancers [110–114]. Furthermore, dextran-based nanogels have been used in PCI applications for siRNA delivery [115]. Lee et al. (2017) fabricated Ce6-loaded gold-stabilized carboxymethyl dextran NPs (Ce6-GS-CNPs) to treat mouse head and neck tumor models *in vivo*. The NPs were found to maintain their highly stable structure and demonstrate no significant changes in their size for 6 days in the presence of serum. When administrated intravenously, the NPs displayed prolonged circulation in the body of mice as well as gradual accumulation in the tumor tissue. Upon exposure to laser irradiation, the tumor site could be visualized through near-infrared fluorescence imaging system and an effective suppression of tumor growth was observed [113]. In the PCI study by Raemdonck et al. (2010), the cationic dextran nanogels were utilized as siRNA depots within the endosomes allowing the siRNA to be released into the cytosol at a desired time with the aid of PCI in order to prolong the knockdown of enhanced green fluorescent protein (EGFP) for up to a week post transfection in hepatoma cells. For the application of PCI, sensitizer TPPS$_{2a}$ was used. A major fraction of internalized cationic nanogels had shown to be trafficked toward acidic vesicles in a previous study by this group [116]. The use of PCI therefore led to the release of a fraction of the endocytosed siEGFP nanogels into the cytosol thus resulting in an extra dose of siEGFP being released into the cytoplasm of the cells and the duration of EGFP knockdown being substantially enhanced. Furthermore, using such method allows RNAi effects to last for a longer period with one single dose of the siRNA nanocarrier which circumvents the need for a second dose application in order to obtain the same effect for the same duration [115].

Although the use of synthetic polymers in drug delivery systems appears more desirable due to their capacity in adjusting their mechanical properties and degradation kinetics to meet the requirements for various applications, natural polymers such as chitosan have attracted wider attention because of their availability, lower costs as well as potential of being chemically modified [117]. Despite the progress made in the field of drug delivery with the aid of synthetic biodegradable polymers, natural polymers still continue to be an area of interest in research.

4.4   NON-BIODEGRADABLE NANOPARTICLES

Due to lack of controlled drug release and degradability, non-biodegradable NPs tend to function in a different manner in PDT. Since such NPs are not prone to destruction by treatment processes, they are suitable for repeated use given that the appropriate activation conditions are present. Non-biodegradable NPs hold several advantages in comparison to biodegradable polymeric NPs. These advantages include: (1) convenient particle size, shape, porosity, and monodispersibility management; (2) the NP's ability to remain stable during environmental fluctuations; (3) the resistance of the NPs toward microbial attacks; and (4) continuous oxygen diffusion through the NPs due to the adjustable pore sizes [117]. Most non-biodegradable NPs are silica or metallic based. The unique properties of metallic NPs have made them subjects of extensive research to investigate their potential applicability in biochemistry as chemical and biological sensors, in nanoelectronics and nanostructured magnetism as systems as well as in medicine as drug delivery tools. In contrast to most silica-based NPs, metallic NPs require photosensitizers to be conjugated to them on their surfaces [118]. Gold (Au) NPs are an example of metallic NPs that have been widely investigated in PDT due to their inert chemical properties as well as the ability to cause minimal acute cytotoxicity [119]. An example of photosensitizer conjugation to AuNPs has been demonstrated in Figure 4.1a.

4.4.1 GOLD NANOPARTICLES

Gold NPs (AuNPs) have several exceptional properties, e.g., the surface plasmon resonance (SPR) effect as well as the capability to transform light energy to heat for PTT applications. Such properties have enabled the development of various photothermal and photodynamic agents [5, 120–129]. Furthermore, the field enhancement of incidence light surrounding AuNPs, could be employed for improving the excitation efficiency of the coloaded photosensitizer due to the presence of the localized surface plasmon resonance [5].

In an interesting study by Chen et al. (2013), gold nanorods (AuNR) and indocyanine green (ICG) (photosensitizer) were integrated into chitosan nanospheres (CS-AuNR-ICG NSs) through a nonsolvent counterion complexation method as well as electrostatic interaction in order to develop a photothermal/photodynamic dual-modality system that could be simultaneously triggered by single near-infrared laser for the treatment of murine hepatic tumors in *in vivo* models. The intravenous injection of CS-AuNR-ICG NSs followed by 10-min irradiation with 808-nm laser 8 hr later, resulted in significant tumor ablation as well as hampering of tumor growth and prolonged lifetime because of the synergistic PTT/PDT effect which made the outcome more superior than that observed with PTT or PDT alone [120].

Nombona et al. (2012), on the other hand, investigated the change in the efficiency of PDT with zinc phthalocyanine (Pc) conjugated to AuNPs or encapsulated in liposomes in breast cancer cells *in vitro*. Upon irradiation with 676-nm laser, the cell viability dropped to 60.1% for Pc-AuNP and 51.9% for liposome bound Pc. Interestingly, compared to PDT with Pc alone, PDT with Pc-AuNP resulted in only a moderate decrease in cell viability [128].

Zhang et al. (2015), however, conjugated 5-aminolevulinic acid to AuNPs to observe PDT efficiency in human chronic myeloid leukemia cells. The cells were incubated with 2 mM 5-ALA-AuNPs conjugates and exposed to 3 different light sources 6 hr later. The light sources

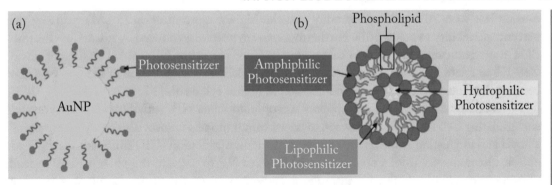

Figure 4.1: Examples of photosensitizer and NP conjugation and encapsulation. (a) Conjugation of photosensitizer to surface of AuNPs. (b) Encapsulation of photosensitizer(s) in different locations of the lipid bilayer of the liposome depending on the properties of the photosensitizers. Hydrophilic, lipophilic, and amphiphilic photosensitizers become localized within the core, hydrophobic tail region, and near the hydrophilic surface of the liposome, respectively.

used were 502-nm LED (1 hr irradiation), 635-nm continuous wavelength laser (10-min irradiation) and xenon lamp (10-min irradiation). Cells that underwent treatment with 5-ALA-AuNP conjugates showed superior cell killing efficiency in comparison to cells treated with 5-ALA only. Cultures treated with AuNPs only demonstrated almost no cellular damage. Out of the three light treatments used, the best cell killing efficiency was achieved with 1 hr irradiation with 502-nm LED array [126].

4.4.2   SILICA-BASED NANOPARTICLES

The most common types of silica-based NPs used in PDT applications include Stöber silica NPs, organically modified silica (ORMOSIL), and mesoporous silica NPs (MSN) [130–143]. Properties associated with these NPs such as chemical inertness, matrix transparency to light absorption in addition to porosity that lacks susceptibility toward swelling, and changes related to pH variation make them promising as vectors in PDT applications. Silica-based NPs can be developed through a variety of synthesis protocols using numerous precursors. Furthermore, these NPs are adaptable with regards to their size, porosity, shape, as well as dispersibility and can be targeted to cancer cells via surface modifications with different functional groups, polymers, and targeting biomolecules [5].

Bharathiraja et al. (2017) formulated Ce6-conjugated and folic acid (FA)-decorated silica NPs (silica-Ce6-FA) utilizing the Stöber method for targeted phosensitizer delivery to breast cancer cells *in vitro* and enhancement of PDT efficiency. The cells were incubated with different concentrations of silica-Ce6-FA for 1 hr followed by 10-min of irradiation with 670-nm laser. The reductions in cell viability post PDT treatment with silica-Ce6-FA were consistent with the increase in NP concentrations, with the lowest concentration of the NP (0.575 uM) applied causing less than 20% cellular mortality and the highest concentration (5 uM) reducing the percentage viability to below 30%. Furthermore, the cytotoxicity induced by silica-Ce6-FA post PDT was significantly higher than that caused by Ce6. The use of FA enabled binding of silica-Ce6-FA to folate receptors which helped to increase cellular accumulation of the phosensitizer compared to free Ce6 application and thus led to higher cell death [130].

Kamkaew et al. (2016) used hollow mesoporous silica NPs (HMSNPs) as a carrier for encapsulating Ce6 for PDT application in breast tumor models *in vivo*. This study incorporated a novel and interesting approach by employing Cerenkov radiation (CR) from radionuclide oxophilic zirconium-89 (89Zr, t1/2 = 78.4 h) in silica NP to activate the phosensitizer instead of using external light. Following intratumoral administration, the nanoconstruct was found to be retained in the tumor area for up to 14 days. CR-induced PDT in the mice led to complete tumor growth inhibition within 14 days after injection. The percentage of tumor size reductions after 14 days in mice treated with [89Zr]HMSN-Ce6, [89Zr]HMSN (radiolabeled NP only), and HMSN-Ce6 (unlabeled phosensitizer carrying NP only) were 75%, 20%, and 32%, respectively, compared to the control group. Furthermore, a majority of the tumor tissue treated with [89Zr]HMSNCe6 were found to be destroyed while most of the tumor tissues in the other three groups retained their normal morphology [134].

Qian et al. (2012) developed organically modified silica (ORMOSIL) NPs encapsulated with either PpIX (protoporphyrin IX) photosensitizers or IR-820 NIR fluorophores for conducting imaging and two photon PDT in cervical cancer *in vitro* and *in vivo* models. The *in vitro* samples underwent incubation with PpIX doped ORMOSIL for 2 hr prior to irradiation with 800-nm laser for a duration of 2-min. The *in vitro* study results showed that PpIX molecules were able to be transferred effectively into the cancer cells using ORMOSIL NPs as carriers. The 2-min irradiation period did not induce any morphological changes in the cells; however, the cells became round and bubbles were formed on their surfaces when the irradiation period was extended to 8-min. The bubbles increased and signs of necrosis appeared around the cells, 15-min post irradiation. The intradermal injection of PEG modified IR-820 doped ORMOSIL NPs into forepaw pad of the mouse enabled rapid diffusion of the NPs from the site of injection into the lymphatics. NIR fluorescence signals were observed in the axillary lymph node (ALN), 4-min post injection with the signal becoming stronger and also visible in the sentinel lymph node (SLN) with increase in time and finally reaching maximum intensity after 20-min. After this point the NPs migrated slowly from the SLN causing the NIR fluorescence signal intensity in the SLN to decrease. Furthermore, NIR signals were observed in the liver, tumor, and tail of the mouse treated intravenously with IR-820 doped ORMOSIL NPs indicating that a majority of the NPs had accumulated in the liver through blood circulation and the NPs were able to avert capture/degradation by the reticuloendothelial system (RES) because of the presence of PEG molecules on their surfaces. However, the accumulation of the NPs in the tumor increased 24 hr post treatment. All NPs were found to have been cleared from the tumor and liver 30 days later causing no changes to the health and behavior of the mouse since the OSMOSIL NPs do not induce organ/tissue cytotoxicity or damage [140].

4.5   REFERENCES

[1] Muehlmann, L. A., Ma, B. C., Longo, J. P., Almeida Santos, M. de F. M., and Azevedo, R. B. 2014. Aluminum-phthalocyanine chloride associated to poly(methyl vinyl ether-co-maleic anhydride) nanoparticles as a new third-generation photosensitizer for anticancer photodynamic therapy, *Int. J. Nanomed.*, 9:1199–1213. DOI: 10.2147/IJN.S57420. 41

[2] Chatterjee, D. K., Fong, L. S., and Zhang. Y. 2008. Nanoparticles in photodynamic therapy: An emerging paradigm, *Adv. Drug Deliv. Rev.*, 60:1627–1637. DOI: 10.1016/j.addr.2008.08.003. 41, 42

[3] Jain, K. K. 2008. Recent advances in nanooncology, *Technol. Cancer Res. Treat.*, 7:1–13. DOI: 10.1177/153303460800700101. 41

[4] Jerjes, W., Theodossiou, T. A., Hirschberg, H., Hogset, A., Weyergang, A., Selbo, P. K., Hamdoon, Z., Hopper, C., and Berg, K. 2020. Photochemical internalization for intracellular drug delivery. From basic mechanisms to clinical research, *J. Clin. Med.*, 9:528. DOI: 10.3390/jcm9020528. 41

[5] Lucky, S. S., Soo, K. C., and Zhang, Y. 2015. Nanoparticles in photodynamic therapy, *Chem. Rev.*, 115:1990–2042. DOI: 10.1021/cr5004198. 41, 42, 45, 46, 47, 48, 52, 53

[6] Mohammad-Hadi, L., MacRobert, A. J., Loizidou, M., and Yaghini, E. 2018. Photodynamic therapy in 3D cancer models and the utilisation of nanodelivery systems, *Nanoscale*, 10:1570–1581. DOI: 10.1039/c7nr07739d. 41

[7] Bamrungsap, S., Zhao, Z., Chen, T., Wang, L., Li, C., Fu, T., and Tan, W. 2012. Nanotechnology in therapeutics: A focus on nanoparticles as a drug delivery system, *Nanomed.*, 7:1253–1271. DOI: 10.2217/nnm.12.87. 41, 42

[8] Bazak, R., Houri, M., Achy, S. E., Hussein, W., and Refaat, T. 2014. Passive targeting of nanoparticles to cancer: A comprehensive review of the literature, *Mol. Clin. Oncol.*, 2:904–908. DOI: 10.3892/mco.2014.356. 41

[9] Sibani, S. A., McCarron, P. A., Woolfson, A. D., and Donnelly, R. F. 2008. Photosensitiser delivery for photodynamic therapy. Part 2: Systemic carrier platforms, *Expert Opin. Drug Deliv.*, 5:1241–1254. DOI: 10.1517/17425240802444673. 41

[10] Firczuk, M., Winiarska, M., Szokalska, A., Jodlowska, M., Swiech, M., Bojarczuk, K., Salwa, P., and Nowis, D. 2011. Approaches to improve photodynamic therapy of cancer, *Front Biosci.*, 16:208–224, Landmark Ed. DOI: 10.2741/3684. 41, 42

[11] Maeda, H., Wu, J., Sawa, T., Matsumura, Y., and Hori, K. 2000. Tumor vascular permeability and the EPR effect in macromolecular therapeutics: A review, *J. Control Release*, 65:271–284. DOI: 10.1016/s0168-3659(99)00248-5. 41

[12] Davis, M. E., Chen, Z. G., and Shin, D. M. 2008. Nanoparticle therapeutics: An emerging treatment modality for cancer, *Nat. Rev. Drug Discov.*, 7:771–782. DOI: 10.1038/nrd2614. 41

[13] Konan-Kouakou, Y. N., Boch, R., Gurny, R., and Allemann, E. 2005. In vitro and in vivo activities of verteporfin-loaded nanoparticles, *J. Control Release*, 103:83–91. DOI: 10.1016/j.jconrel.2004.11.023. 41

[14] Master, A., Livingston, M., and Sen Gupta, A. 2013. Photodynamic nanomedicine in the treatment of solid tumors: Perspectives and challenges, *J. Control Release*, 168:88–102. DOI: 10.1016/j.jconrel.2013.02.020. 42

[15] Allison, R. R., Bagnato, V. S., and Sibata, C. H. 2010. Fut. of oncologic photodynamic therapy, *Fut. Oncol.*, 6:929–940. DOI: 10.2217/fon.10.51. 42

[16] Konan, Y. N., Gurny, R., and Allemann, E. 2002. State of the art in the delivery of photosensitizers for photodynamic therapy, *J. Photochem. Photobiol. B*, 66:89–106. DOI: 10.1016/s1011-1344(01)00267-6. 42

[17] Yin, R., Wang, M., Huang, Y. Y., Huang, H. C., Avci, P., Chiang, L. Y., and Hamblin, M. R. 2014. Photodynamic therapy with decacationic [60]fullerene monoadducts: Effect of a light absorbing electron-donor antenna and micellar formulation, *Nanomed.*, 10:795–808. DOI: 10.1016/j.nano.2013.11.014. 42

[18] Bakry, R., Vallant, R. M., Najam-ul-Haq, M., Rainer, M., Szabo, Z., Huck, C. W., and Bonn, G. K. 2007. Medicinal applications of fullerenes, *Int. J. Nanomed.*, 2:639–649. 42

[19] Mroz, P., Tegos, G. P., Gali, H., Wharton, T., Sarna, T., and Hamblin, M. R. 2007. Photodynamic therapy with fullerenes, *Photochem. Photobiol. Sci.*, 6:1139–1149. DOI: 10.1039/b711141j. 42

[20] Tabata, Y., Murakami, Y., and Ikada, Y. 1997. Photodynamic effect of polyethylene glycol-modified fullerene on tumor, *Jpn. J. Cancer Res.*, 88:1108–1116. DOI: 10.1111/j.1349-7006.1997.tb00336.x. 42

[21] Liu, J., Ohta, S., Sonoda, A., Yamada, M., Yamamoto, M., Nitta, N., Murata, K., and Tabata, Y. 2007. Preparation of PEG-conjugated fullerene containing Gd3+ ions for photodynamic therapy, *J. Control Release*, 117:104–110. DOI: 10.1016/j.jconrel.2006.10.008. 42

[22] Guan, M., Ge, J., Wu, J., Zhang, G., Chen, D., Zhang, W., Zhang, Y., Zou, T., Zhen, M., Wang, C., Chu, T., Hao, X., and Shu, C. 2016. Fullerene/photosensitizer nanovesicles as highly efficient and clearable phototheranostics with enhanced tumor accumulation for cancer therapy, *Biomaterials*, 103:75–85. DOI: 10.1016/j.biomaterials.2016.06.023. 42, 43

[23] Aqel, A., Yusuf, K., Al-Othman, Z. A., Badjah-Hadj-Ahmed, A. Y., and Alwarthan, A. A. 2012. Effect of multi-walled carbon nanotubes incorporation into benzyl methacrylate monolithic columns in capillary liquid chromatography, *Analyst*, 137:4309–4317. DOI: 10.1039/c2an35518c. 43

[24] Madani, S. Y., Naderi, N., Dissanayake, O., Tan, A., and Seifalian, A. M. 2011. A new era of cancer treatment: Carbon nanotubes as drug delivery tools. *Int. J. Nanomed.*, 6:2963–2979. DOI: 10.2147/IJN.S16923 . 43

[25] Zhu, W., Duan, H., and Bolton, K. 2009. Diameter and chirality changes of single-walled carbon nanotubes during growth: An ab-initio study, *J. Nanosci. Nanotechnol.*, 9:1222–1225. DOI: 10.1166/jnn.2009.c124. 43

[26] Wang, L., Shi, J., Liu, R., Liu, Y., Zhang, J., Yu, X., Gao, J., Zhang, C., and Zhang, Z. 2014. Photodynamic effect of functionalized single-walled carbon nanotubes: A potential sensitizer for photodynamic therapy, *Nanoscale*, 6:4642–4651. DOI: 10.1039/c3nr06835h. 43

[27] Zhang, P., Huang, H., Huang, J., Chen, H., Wang, J., Qiu, K., Zhao, D., Ji, L., and Chao, H. 2015. Noncovalent ruthenium(II) complexes-single-walled carbon nanotube composites for bimodal photothermal and photodynamic therapy with near-infrared irradiation, *ACS Appl. Mater. Interf.*, 7:23278–23290. DOI: 10.1021/acsami.5b07510. 43

[28] Shao, L., Gao, Y., and Yan, F. 2011. Semiconductor quantum dots for biomedicial applications, *Sensors (Basel)*, 11:11736–11751. DOI: 10.3390/s111211736. 43

[29] Misra, R., Acharya, S., and Sahoo, S. K. 2010. Cancer nanotechnology: Application of nanotechnology in cancer therapy. *Drug Discov. Today*, 15:842–850. DOI: 10.1016/j.drudis.2010.08.006. 43

[30] Juzenas, P., Chen, W., Sun, Y. P., Coelho, M. A., Generalov, R., Generalova, N., and Christensen, I. L. 2008. Quantum dots and nanoparticles for photodynamic and radiation therapies of cancer, *Adv. Drug Deliv. Rev.*, 60:1600–1614. DOI: 10.1016/j.addr.2008.08.004. 43

[31] Thomas, A., Nair, P. V., and Thomas, K. G. 2014. InP quantum dots: An environmentally friendly material with resonance energy transfer requisites, *J. Phys. Chem. C*, 118:3838–3845. DOI: 10.1021/jp500125v. 43

[32] Yaghini, E., Turner, H. D., Le Marois, A. M., Suhling, K., Naasani, I., and MacRobert, A. J. 2016. In vivo biodistribution studies and ex vivo lymph node imaging using heavy metal-free quantum dots, *Biomaterials*, 104:182–191. DOI: 10.1016/j.biomaterials.2016.07.014. 43

[33] Yaghini, E., Seifalian, A. M., and MacRobert, A. J. 2009. Quantum dots and their potential biomedical applications in photosensitization for photodynamic therapy, *Nanomedicine (Lond)*, 4:353–363. DOI: 10.2217/nnm.09.9. 43

[34] Qi, Z. D., Li, D. W., Jiang, P., Jiang, F. L., Li, Y. S., Liu, Y., Wong, W. K., and Cheah, K. W. 2011. Biocompatible CdSe quantum dot-based photosensitizer under two-photon excitation for photodynamic therapy, *J. Mater. Chem.*, 21:2455–2458. DOI: 10.1039/c0jm03229h. 43

[35] Liu, Y., Xu, Y., Geng, X., Huo, Y., Chen, D., Sun, K., Zhou, G., Chen, B., and Tao, K. 2018. Synergistic targeting and efficient photodynamic therapy based on graphene oxide quantum dot-upconversion nanocrystal hybrid nanoparticles, *Small*, 14:e1800293. DOI: 10.1002/smll.201800293. 43

[36] Ge, J., Lan, M., Zhou, B., Liu, W., Guo, L., Wang, H., Jia, Q., Niu, G., Huang, X., Zhou, H., Meng, X., Wang, P., Lee, C. S., Zhang, W., and Han, X. 2014. A graphene quantum dot photodynamic therapy agent with high singlet oxygen generation, *Nat. Commun.*, 5:4596. DOI: 10.1038/ncomms5596. 43, 44

[37] Monroe, J. D., Belekov, E., Er, A. O., and Smith, M. E. 2019. Anticancer photodynamic therapy properties of sulfur-doped graphene quantum dot and methylene blue preparations in MCF-7 breast cancer cell culture, *Photochem. Photobiol.*, 95:1473–1481. DOI: 10.1111/php.13136. 43

[38] Wang, C., Cheng, L., and Liu, Z. 2011. Upconversion nanoparticles for potential cancer theranostics, *Ther. Deliv.*, 2:1235–1239. DOI: 10.4155/tde.11.93. 44

[39] Jin, J., Xu, Z., Zhang, Y., Gu, Y. J., Lam, M. H., and Wong, W. T. 2013. Upconversion nanoparticles conjugated with Gd(3+)—DOTA and RGD for targeted dual-modality imaging of brain tumor xenografts, *Adv. Healthc. Mater.*, 2:1501–1512. DOI: 10.1002/adhm.201300102. 44

[40] Wang, C., Tao, H., Cheng, L., and Liu, Z. 2011. Near-infrared light induced in vivo photodynamic therapy of cancer based on upconversion nanoparticles, *Biomaterials*, 32:6145–6154. DOI: 10.1016/j.biomaterials.2011.05.007. 44

[41] Idris, N. M., Gnanasammandhan, M. K., Zhang, J., Ho, P. C., Mahendran, R., and Zhang, Y. 2012. In vivo photodynamic therapy using upconversion nanoparticles as

remote-controlled nanotransducers, *Nat. Med.*, 18:1580–1585. DOI: 10.1038/nm.2933. 44

[42] Guo, H., Qian, H., Idris, N. M., and Zhang, Y. 2010. Singlet oxygen-induced apoptosis of cancer cells using upconversion fluorescent nanoparticles as a carrier of photosensitizer, *Nanomed. Nanotechnol., Biol. Med.*, 6:486–495. DOI: 10.1016/j.nano.2009.11.004. 44

[43] Wang, M., Chen, Z., Zheng, W., Zhu, H., Lu, S., Ma, E., Tu, D., Zhou, S., Huang, M., and Chen, X. 2014. Lanthanide-doped upconversion nanoparticles electrostatically coupled with photosensitizers for near-infrared-triggered photodynamic therapy, *Nanoscale*, 6:8274–8282. DOI: 10.1039/c4nr01826e. 44

[44] Ang, L. Y., Lim, M. E., Ong, L. C., and Zhang, Y. 2011. Applications of upconversion nanoparticles in imaging, detection and therapy, *Nanomedicine (Lond)*, 6:1273–1288. DOI: 10.2217/nnm.11.108. 44

[45] Zhang, Z., Jayakumar, M. K. G., Zheng, X., Shikha, S., Zhang, Y., Bansal, A., Poon, D. J. J., Chu, P. L., Yeo, E. L. L., Chua, M. L. K., Chee, S. K., and Zhang, Y. 2019. Upconversion superballs for programmable photoactivation of therapeutics, *Nat. Commun.*, 10:4586. DOI: 10.1038/s41467-019-12506-w. 44

[46] Zhu, K., Liu, G., Hu, J., and Liu, S. 2017. Near-infrared light-activated photochemical internalization of reduction-responsive polyprodrug vesicles for synergistic photodynamic therapy and chemotherapy, *Biomacromolecules*, 18:2571–2582. DOI: 10.1021/acs.biomac.7b00693. 44

[47] Jayakumar, M. K., Bansal, A., Huang, K., Yao, R., Li, B. N., and Zhang, Y. 2014. Near-infrared-light-based nano-platform boosts endosomal escape and controls gene knockdown in vivo, *ACS Nano*, 8:4848–4858. DOI: 10.1021/nn500777n. 44

[48] Li, L. and Huh, K. M. 2014. Polymeric nanocarrier systems for photodynamic therapy, *Biomater. Res.*, 18:19. DOI: 10.1186/2055-7124-18-19. 45

[49] Paszko, E., Ehrhardt, C., Senge, M. O., Kelleher, D. P., and Reynolds, J. V. 2011. Nanodrug applications in photodynamic therapy, *Photodiagn. Photodyn. Ther.*, 8:14–29. DOI: 10.1016/j.pdpdt.2010.12.001. 45

[50] Syu, W. J., Yu, H. P., Hsu, C. Y., Rajan, Y. C., Hsu, Y. H., Chang, Y. C., Hsieh, W. Y., Wang, C. H., and Lai, P. S. 2012. Improved photodynamic cancer treatment by folate-conjugated polymeric micelles in a KB xenografted animal model, *Small*, 8:2060–2069. DOI: 10.1002/smll.201102695. 45

[51] Peng, C. L., Shieh, M. J., Tsai, M. H., Chang, C. C., and Lai, P. S. 2008. Self-assembled star-shaped chlorin-core poly(epsilon-caprolactone)-poly(ethylene glycol) diblock copolymer micelles for dual chemo-photodynamic therapies, *Biomaterials*, 29:3599–3608. DOI: 10.1016/j.biomaterials.2008.05.018. 45

[52] Ding, H., Yu, H., Dong, Y., Tian, R., Huang, G., Boothman, D. A., Sumer, B. D., and Gao, J. 2011. Photoactivation switch from type II to type I reactions by electron-rich micelles for improved photodynamic therapy of cancer cells under hypoxia, *J. Control Release*, 156:276–280. DOI: 10.1016/j.jconrel.2011.08.019. 45

[53] Konan, Y. N., Berton, M., Gurny, R., and Allémann, E. 2003. Enhanced photodynamic activity of meso-tetra(4-hydroxyphenyl)porphyrin by incorporation into sub-200 nm nanoparticles, *Eur. J. Pharm. Sci.*, 18:241–249. DOI: 10.1016/s0928-0987(03)00017-4. 45

[54] McCarthy, J. R., Perez, J. M., Brückner, C., and Weissleder, R. 2005. Polymeric nanoparticle preparation that eradicates tumors, *Nano Lett.*, 5:2552–2556. DOI: 10.1021/nl0519229. 45

[55] Hung, H. I., Klein, O. J., Peterson, S. W., Rokosh, S. R., Osseiran, S., Nowell, N. H., and Evans, C. L. 2016. PLGA nanoparticle encapsulation reduces toxicity while retaining the therapeutic efficacy of EtNBS-PDT in vitro, *Sci. Rep.*, 6:33234. DOI: 10.1038/srep33234. 45

[56] Boeuf-Muraille, G., Rigaux, G., Callewaert, M., Zambrano, N., Van Gulick, L., Roullin, V. G., Terryn, C., Andry, M. C., Chuburu, F., Dukic, S., and Molinari, M. 2019. Evaluation of mTHPC-loaded PLGA nanoparticles for in vitro photodynamic therapy on C6 glioma cell line, *Photodiagn. Photodyn. Ther.*, 25:448–455. DOI: 10.1016/j.pdpdt.2019.01.026. 45

[57] da Silva, C. L., Del Ciampo, J. O., Rossetti, F. C., Bentley, M. V., and Pierre, M. B. 2013. PLGA nanoparticles as delivery systems for protoporphyrin IX in topical PDT: Cutaneous penetration of photosensitizer observed by fluorescence microscopy, *J. Nanosci. Nanotechnol.*, 13:6533–6540. DOI: 10.1166/jnn.2013.7789. 45

[58] Tian, J., Xu, L., Xue, Y., Jiang, X., and Zhang, W. 2017. Enhancing photochemical internalization of DOX through a porphyrin-based amphiphilic block copolymer, *Biomacromolecules*, 18:3992–4001. DOI: 10.1021/acs.biomac.7b01037. 45, 46

[59] Park, H., Park, W., and Na, K. 2014. Doxorubicin loaded singlet-oxygen producible polymeric micelle based on chlorine e6 conjugated pluronic F127 for overcoming drug resistance in cancer, *Biomaterials*, 35:7963–7969. DOI: 10.1016/j.biomaterials.2014.05.063. 45, 46

[60] Lu, H. L., Syu, W. J., Nishiyama, N., Kataoka, K., and Lai, P. S. 2011. Dendrimer phthalocyanine-encapsulated polymeric micelle-mediated photochemical internalization extends the efficacy of photodynamic therapy and overcomes drug-resistance in vivo, *J. Control Release*, 155:458–464. DOI: 10.1016/j.jconrel.2011.06.005. 45

[61] Yen, H. C., Cabral, H., Mi, P., Toh, K., Matsumoto, Y., Liu, X., Koori, H., Kim, A., Miyazaki, K., Miura, Y., Nishiyama, N., and Kataoka, K. 2014. Light-induced cytosolic activation of reduction-sensitive camptothecin-loaded polymeric micelles for spatiotemporally controlled in vivo chemotherapy, *ACS Nano*, 8:11591–11602. DOI: 10.1021/nn504836s. 45

[62] Gargouri, M., Sapin, A., Arica-Yegin, B., Merlin, J. L., Becuwe, P., and Maincent, P. 2011. Photochemical internalization for pDNA transfection: Evaluation of poly(d,l-lactide-co-glycolide) and poly(ethylenimine) nanoparticles, *Int. J. Pharm.*, 403:276–284. DOI: 10.1016/j.ijpharm.2010.10.040. 45

[63] Xiao, Z., Hansen, C. B., Allen, T. M., Miller, G. G., and Moore, R. B. 2005. Distribution of photosensitizers in bladder cancer spheroids: Implications for intravesical instillation of photosensitizers for photodynamic therapy of bladder cancer, *J. Pharm. Pharm. Sci.* 8:536–543. 46

[64] Gaio, E., Scheglmann, D., Reddi, E., and Moret, F. 2016. Uptake and photo-toxicity of Foscan®, Foslip® and Fospeg® in multicellular tumor spheroids, *J. Photochem. Photobiol. B*, 161:244–252. DOI: 10.1016/j.jphotobiol.2016.05.011. 46

[65] Lee, J., Kim, J., Jeong, M., Lee, H., Goh, U., Kim, H., Kim, B., and Park, J. H. 2015. Liposome-based engineering of cells to package hydrophobic compounds in membrane vesicles for tumor penetration, *Nano Lett.*, 15:2938–2944. DOI: 10.1021/nl5047494. 46

[66] Yang, Y., Wang, L., Cao, H., Li, Q., Li, Y., Han, M., Wang, H., and Li, J. 2019. Photodynamic therapy with liposomes encapsulating photosensitizers with aggregation-induced emission, *Nano Lett.*, 19:1821–1826. DOI: 10.1021/acs.nanolett.8b04875. 46

[67] Reshetov, V., Lassalle, H. P., Francois, A., Dumas, D., Hupont, S., Grafe, S., Filipe, V., Jiskoot, W., Guillemin, F., Zorin, V., and Bezdetnaya, L. 2013. Photodynamic therapy with conventional and PEGylated liposomal formulations of mTHPC (temoporfin): Comparison of treatment efficacy and distribution characteristics in vivo, *Int. J. Nanomed.*, 8:3817–3831. DOI: 10.2147/ijn.s51002. 46

[68] Lovell, J. F., Jin, C. S., Huynh, E., Jin, H., Kim, C., Rubinstein, J. L., Chan, W. C., Cao, W., Wang, L. V., and Zheng, G. 2011. Porphysome nanovesicles generated by porphyrin bilayers for use as multimodal biophotonic contrast agents, *Nat. Mater.*, 10:324–332. DOI: 10.1038/nmat2986. 47

[69] Galanzha, E. I., Kokoska, M. S., Shashkov, E. V., Kim, J. W., Tuchin, V. V., and Zharov, V. P. 2009. In vivo fiber-based multicolor photoacoustic detection and phototothermal purging of metastasis in sentinel lymph nodes targeted by nanoparticles, *J. Biophot.*, 2:528–539. DOI: 10.1002/jbio.200910046. 47

[70] Yaghini, E., Dondi, R., Edler, K. J., Loizidou, M., MacRobert, A. J., and Eggleston, I. M. 2018. Codelivery of a cytotoxin and photosensitiser via a liposomal nanocarrier: A novel strategy for light-triggered cytosolic release, *Nanoscale*, 10:20366–20376. DOI: 10.1039/c8nr04048f. 47

[71] Galanou, M. C., Theodossiou, T. A., Tsiourvas, D., Sideratou, Z., and Paleos, C. M. 2008. Interactive transport, subcellular relocation and enhanced phototoxicity of hypericin encapsulated in guanidinylated liposomes via molecular recognition, *Photochem. Photobiol.*, 84:1073–1083. DOI: 10.1111/j.1751-1097.2008.00392.x. 47

[72] Hellum, M., Hogset, A., Engesaeter, B. O., Prasmickaite, L., Stokke, T., Wheeler, C., and Berg, K. 2003. Photochemically enhanced gene delivery with cationic lipid formulations, *Photochem. Photobiol. Sci.*, 2:407–411. DOI: 10.1039/b211880g. 47

[73] Svenson, S. and Tomalia, D. A. 2005. Dendrimers in biomedical applications—reflections on the field, *Adv. Drug Deliv. Rev.*, 57:2106–2129. DOI: 10.1016/j.addr.2005.09.018. 47

[74] Avci, P., Erdem, S. S., and Hamblin, M. R. 2014. Photodynamic therapy: One step ahead with self-assembled nanoparticles, *J. Biomed. Nanotechnol.*, 10:1937–1952. DOI: 10.1166/jbn.2014.1953. 47, 48

[75] Boas, U. and Heegaard, P. M. 2004. Dendrimers in drug research, *Chem. Soc. Rev.*, 33:43–63. DOI: 10.1039/b309043b. 47

[76] Wolinsky, J. B. and Grinstaff, M. W. 2008. Therapeutic and diagnostic applications of dendrimers for cancer treatment, *Adv. Drug Deliv. Rev.*, 60:1037–1055. DOI: 10.1016/j.addr.2008.02.012. 47

[77] Battah, S., O'Neill, S., Edwards, C., Balaratnam, S., Dobbin, P., and MacRobert, A. J. 2006. Enhanced porphyrin accumulation using dendritic derivatives of 5-aminolaevulinic acid for photodynamic therapy: An in vitro study, *Int. J. Biochem. Cell Biol.*, 38:1382–1392. DOI: 10.1016/j.biocel.2006.02.001. 48

[78] Di Venosa, G. M., Casas, A. G., Battah, S., Dobbin, P., Fukuda, H., Macrobert, A. J., and Batlle, A. 2006. Investigation of a novel dendritic derivative of 5-aminolaevulinic acid for photodynamic therapy, *Int. J. Biochem. Cell Biol.*, 38:82–91. DOI: 10.1016/j.biocel.2005.08.001. 48

[79] Battah, S., Balaratnam, S., Casas, A., O'Neill, S., Edwards, C., Batlle, A., Dobbin, P., and MacRobert, A. J. 2007. Macromolecular delivery of 5-aminolaevulinic acid for photodynamic therapy using dendrimer conjugates, *Molec. Cancer Therapeut.*, 6:876–885. DOI: 10.1158/1535-7163.mct-06-0359. 48

[80] Casas, A., Battah, S., Di Venosa, G., Dobbin, P., Rodriguez, L., Fukuda, H., Batlle, A., and MacRobert, A. J. 2009. Sustained and efficient porphyrin generation in vivo using dendrimer conjugates of 5-ALA for photodynamic therapy, *J. Control Release*, 135:136–143. DOI: 10.1016/j.jconrel.2009.01.002. 48

[81] Karthikeyan, K., Babu, A., Kim, S. J., Murugesan, R., and Jeyasubramanian, K. 2011. Enhanced photodynamic efficacy and efficient delivery of Rose Bengal using nanostructured poly(amidoamine) dendrimers: Potential application in photodynamic therapy of cancer *Cancer Nanotechnol.*, 2:95–103. DOI: 10.1007/s12645-011-0019-3. 48

[82] Militello, M. P., Hernandez-Ramirez, R. E., Lijanova, I. V., Previtali, C. M., Bertolotti, S. G., and Arbeloa, E. M. 2018. Novel PAMAM dendrimers with porphyrin core as potential photosensitizers for PDT applications, *J. Photochem. Photobio. A Chem.*, 353:71–76. DOI: 10.1016/j.jphotochem.2017.10.057. 48

[83] Narsireddy, A., Vijayashree, K., Adimoolam, M. G., Manorama, S. V., and Rao, N. M. 2015. Photosensitizer and peptide-conjugated PAMAM dendrimer for targeted in vivo photodynamic therapy, *Int. J. Nanomed.*, 10:6865–6878. DOI: 10.2147/IJN.S89474. 48

[84] Bøe, S. L., Jorgensen, J. A. L., Longva, A. S., Lavelle, T., Sæbøe-Larssen, S., and Hovig, E. 2013. Light-controlled modulation of gene expression using polyamidoamine formulations, *Nucleic Acid Ther.*, 23:160–165. DOI: 10.1089/nat.2012.0413. 48

[85] Yuan, A., Hu, Y., and Ming, X. 2015. Dendrimer conjugates for light-activated delivery of antisense oligonucleotides, *RSC Adv.*, 5:35195–35200. DOI: 10.1039/c5ra04091d. 48

[86] Shieh, M. J., Peng, C. L., Lou, P. J., Chiu, C. H., Tsai, T. Y., Hsu, C. Y., Yeh, C. Y., and Lai, P. S. 2008. Non-toxic phototriggered gene transfection by PAMAM-porphyrin conjugates, *J. Control Release*, 129:200–206. DOI: 10.1016/j.jconrel.2008.03.024. 48, 49

[87] Preuß, A., Chen, K., Hackbarth, S., Wacker, M., Langer, K., and Röder, B. 2011. Photosensitizer loaded HSA nanoparticles II: In vitro investigations, *Pharmaceut. Nanotechnol.*, 404:308–316. DOI: 10.1016/j.ijpharm.2010.11.023. 49

[88] Jeong, H., Huh, M., Lee, S. J., Koo, H., Kwon, I. C., Jeong, S. Y., and Kim, K. 2011. Photosensitizer-conjugated human serum albumin nanoparticles for effective photodynamic therapy, *Theranostics*, 1:230–239. DOI: 10.7150/thno/v01p0230. 49, 50

[89] Jiang, C., Cheng, H., Yuan, A., Tang, X., Wu, J., and Hu, Y. 2015. Hydrophobic IR780 encapsulated in biodegradable human serum albumin nanoparticles for photothermal and photodynamic therapy, *Acta Biomater.*, 14:61–69. DOI: 10.1016/j.actbio.2014.11.041. 49

[90] Qu, N., Sun, Y., Li, Y., Hao, F., Qiu, P., Teng, L., Xie, J., and Gao, Y. 2019. Docetaxel-loaded human serum albumin (HSA) nanoparticles: Synthesis, characterization, and evaluation, *Biomed. Eng. Online*, 18:11. DOI: 10.1186/s12938-019-0624-7. 49

[91] Elzoghby, A. O., Samy, W. M., and Elgindy, N. A. 2012. Albumin-based nanoparticles as potential controlled release drug delivery systems, *J. Control Release*, 157:168–182. DOI: 10.1016/j.jconrel.2011.07.031. 49

[92] Gradishar, W. J. 2006. Albumin-bound paclitaxel: A next-generation taxane, *Expert Opin. Pharmacother.*, 7:1041–1053. DOI: 10.1517/14656566.7.8.1041. 49

[93] Miele, E., Spinelli, G. P., Miele, E., Tomao, F., and Tomao, S. 2019. Albumin-bound formulation of paclitaxel (Abraxane® ABI-007) in the treatment of breast cancer, *Int. J. Nanomed.*, 4:99–105. 49

[94] Kratz, F. 2008. Albumin as a drug carrier: Design of prodrugs, drug conjugates and nanoparticles, *J. Control Release*, 132:171–183. DOI: 10.1016/j.jconrel.2008.05.010. 49

[95] Hejazi, R. and Amiji, M. 2003. Chitosan-based gastrointestinal delivery systems, *J. Control Release*, 89:151–165. DOI: 10.1016/s0168-3659(03)00126-3. 50

[96] Naskar, S., Kuotsu, K., and Sharma, S. 2019. Chitosan-based nanoparticles as drug delivery systems: A review on two decades of research, *J. Drug Target.*, 27:379–393. DOI: 10.1080/1061186x.2018.1512112. 50

[97] Kim, J. H., Kim, Y. S., Park, K., Kang, E., Lee, S., Nam, H. Y., Kim, K., Park, J. H., Chi, D. Y., Park, R. W., Kim, I. S., Choi, K., and Kwon, I. C. 2008. Self-assembled glycol chitosan nanoparticles for the sustained and prolonged delivery of antiangiogenic small peptide drugs in cancer therapy, *Biomaterials*, 29:1920–1930. DOI: 10.1016/j.biomaterials.2007.12.038. 50

[98] Kamath, P. R. and Sunil, D. 2017. Nano-chitosan particles in anticancer drug delivery: An up-to-date review, *Mini Rev. Med. Chem.*, 17:1457–1487. DOI: 10.2174/1389557517666170228105731. 50

[99] Ghaz-Jahanian, M. A., Abbaspour-Aghdam, F., Anarjan, N., Berenjian, A., and Jafarizadeh-Malmiri, H. 2015. Application of chitosan-based nanocarriers in tumor-targeted drug delivery, *Mol. Biotechnol.*, 57:201–218. DOI: 10.1007/s12033-014-9816-3. 50

[100] Pandya, A. D., Overbye, A., Sahariah, P., Gaware, V. S., Hogset, H., Masson, M., Hogset, A., Maelandsmo, G. M., Skotland, T., Sandvig, K., and Iversen, T. G. 2020. Drug-loaded photosensitizer-chitosan nanoparticles for combinatorial chemo- and photodynamic-therapy of cancer, *Biomacromolecules*, 21:1489–1498. DOI: 10.1021/acs.biomac.0c00061. 50

[101] Ding, Y. F., Li, S., Liang, L., Huang, Q., Yuwen, L., Yang, W., Wang, R., and Wang, L. H. 2018. Highly biocompatible chlorin e6-loaded chitosan nanoparticles for improved photodynamic cancer therapy, *ACS Appl. Mater. Interf.*, 10:9980–9987. DOI: 10.1021/acsami.8b01522. 50

[102] Sun, Y., Chen, Z. L., Yang, X. X., Huang, P., Zhou, X. P., and Du, X. X. 2009. Magnetic chitosan nanoparticles as a drug delivery system for targeting photodynamic therapy, *Nanotechnology*, 20:135102. DOI: 10.1088/0957-4484/20/13/135102. 50

[103] Gaware, V. S., Hakerud, M., Juzeniene, A., Hogset, A., Berg, K., and Masson, M. 2017. Endosome targeting meso-tetraphenylchlorin-chitosan nanoconjugates for photochemical internalization, *Biomacromolecules*, 18:1108–1126. DOI: 10.1021/acs.biomac.6b01670. 50

[104] Lee, S. J., Park, K., Oh, Y. K., Kwon, S. H., Her, S., Kim, I. S., Choi, K., Lee, S. J., Kim, H., Lee, S. G., Kim, K., and Kwon, I. C. 2009. Tumor specificity and therapeutic efficacy of photosensitizer-encapsulated glycol chitosan-based nanoparticles in tumor-bearing mice, *Biomaterials*, 30:2929–2939. DOI: 10.1016/j.biomaterials.2009.01.058. 50

[105] Chung, C. W., Chung, K. D., Jeong, Y. I., and Kang, D. H. 2013. 5-aminolevulinic acid-incorporated nanoparticles of methoxy poly(ethylene glycol)-chitosan copolymer for photodynamic therapy, *Int. J. Nanomed.*, 8:809–819. DOI: 10.2147/ijn.s39615. 50

[106] Lee, S. J., Koo, H., Lee, D. E., Min, S., Lee, S., Chen, X., Choi, Y., Leary, J. F., Park, K., Jeong, S. Y., Kwon, I. C., Kim, K., and Choi, K. 2011. Tumor-homing photosensitizer-conjugated glycol chitosan nanoparticles for synchronous photodynamic imaging and therapy based on cellular on/off system, *Biomaterials*, 32:4021–4029. DOI: 10.1016/j.biomaterials.2011.02.009. 50

[107] Oh, I. H., Min, H. S., Li, L., Tran, T. H., Lee, Y. K., Kwon, I. C., Choi, K., Kim, K., and Huh, K. M. 2013. Cancer cell-specific photoactivity of pheophorbide a-glycol chitosan nanoparticles for photodynamic therapy in tumor-bearing mice, *Biomaterials*, 34:6454–6463. DOI: 10.1016/j.biomaterials.2013.05.017. 50

[108] Marotta, D. E., Cao, W., Wileyto, E. P., Li, H., Corbin, I., Rickter, E., Glickson, J. D., Chance, B., Zheng, G., and Busch, T. M. 2011. Evaluation of bacteriochlorophyll-reconstituted low-density lipoprotein nanoparticles for photodynamic therapy efficacy in vivo, *Nanomedicine (Lond)*, 6:475–487. DOI: 10.2217/nnm.11.8. 50

[109] Jin, H., Lovell, J. F., Chen, J., Ng, K., Cao, W., Ding, L., Zhang, Z., and Zheng, G. 2011. Cytosolic delivery of LDL nanoparticle cargo using photochemical internalization, *Photochem. Photobiol. Sci.*, 10:810–816. DOI: 10.1039/c0pp00350f. 51

[110] Predescu, A. M., Matei, E., Berbecaru, A. C., Pantilimon, C., Dragan, C., Vidu, R., Predescu, C., and Kuncser, V. 2018. Synthesis and characterization of dextran-coated iron oxide nanoparticles, *R. Soc. Open Sci.*, 5:171525. DOI: 10.1098/rsos.171525. 51

[111] Yan, L., Luo, L., Amirshaghaghi, A., Miller, J., Meng, C., You, T., Busch, T. M., Tsourkas, A., and Cheng, Z. 2019. Dextran-benzoporphyrin derivative (BPD) coated superparamagnetic iron oxide nanoparticle (SPION) micelles for T2-weighted magnetic resonance imaging and photodynamic therapy, *Bioconjug. Chem.*, 30:2974–2981. DOI: 10.1021/acs.bioconjchem.9b00676. 51

[112] Ding, Z., Liu, P., Hu, D., Sheng, Z., Yi, H., Gao, G., Wu, Y., Zhang, P., Ling, S., and Cai, L. 2017. Redox-responsive dextran based theranostic nanoparticles for near-infrared/magnetic resonance imaging and magnetically targeted photodynamic therapy, *Biomater. Sci.*, 5:762–771. DOI: 10.1039/c6bm00846a. 51

[113] Lee, M., Lee, H., Vijayakameswara Rao, N., Han, H. S., Jeon, S., Jeon, J., Lee, S., Kwon, S., Suh, Y. D., and Park, J. H. 2017. Gold-stabilized carboxymethyl dextran nanoparticles for image-guided photodynamic therapy of cancer, *J. Mater. Chem. B*, 5:7319–7327. DOI: 10.1039/c7tb01099k. 51

[114] Mbakidi, J. P., Bregier, F., Ouk, T. S., Granet, R., Alves, S., Rivière, E., Chevreux, S., Lemercier, G., and Sol, V. 2015. Magnetic dextran nanoparticles that bear hydrophilic porphyrin derivatives: Bimodal agents for potential application in photodynamic therapy, *ChemPlusChem*, 80:1416–1426. DOI: 10.1002/cplu.201500087. 51

[115] Raemdonck, K., Naeye, B., Hogset, A., Demeester, J., and De Smedt, S. C. 2010. Prolonged gene silencing by combining siRNA nanogels and photochemical internalization, *J. Control Release*, 145:281–288. DOI: 10.1016/j.jconrel.2010.04.012. 51

[116] Raemdonck, K., Naeye, B., Buyens, K., Vandenbroucke, R. E., Hogset, A., Demeester, J., and De Smedt, S. C. 2009. Biodegradable dextran nanogels for RNA interference: Focusing on endosomal escape and intracellular siRNA delivery, *Adv. Funct. Mater.*, 19:1406–1415. DOI: 10.1002/adfm.200801795. 51

[117] Debele, T. A., Peng, S., and Tsai, H. C. 2015. Drug carrier for photodynamic cancer therapy, *Int. J. Mol. Sci.*, 16:22094–22136. DOI: 10.3390/ijms160922094. 51, 52

[118] Wang, F., Li, C., Cheng, J., and Yuan, Z. 2016. Recent advances on inorganic nanoparticle-based cancer therapeutic agents, *Int. J. Environ. Res. Pub. Health*, 13:1182. DOI: 10.3390/ijerph13121182. 52

[119] Zhang, X. 2015. Gold nanoparticles: Recent advances in the biomedical applications, *Cell Biochem. Biophys.*, 72:771–775. DOI: 10.1007/s12013-015-0529-4. 52

[120] Chen, R., Wang, X., Yao, X., Zheng, X., Wang, J., and Jiang, X. 2013. Near-IR-triggered photothermal/photodynamic dual-modality therapy system via chitosan hybrid nanospheres, *Biomaterials*, 34:8314–8322. DOI: 10.1016/j.biomaterials.2013.07.034. 52

[121] El-Sayed, I. H., Huang, X., and El-Sayed, M. A. 2006. Selective laser photo-thermal therapy of epithelial carcinoma using anti-EGFR antibody conjugated gold nanoparticles, *Cancer Lett.*, 239:129–135. DOI: 10.1016/j.canlet.2005.07.035. 52

[122] Van de Broek, B., Devoogdt, N., D'Hollander, A., Gijs, H. L., Jans, K., Lagae, L., Muyldermans, S., Maes, G., and Borghs, G. 2011. Specific cell targeting with nanobody conjugated branched gold nanoparticles for photothermal therapy, *ACS Nano*, 5:4319–4328. DOI: 10.1021/nn1023363. 52

[123] Cheng, X., Sun, R., Yin, L., Chai, Z., Shi, H., and Gao, M. 2017. Light-triggered assembly of gold nanoparticles for photothermal therapy and photoacoustic imaging of tumors in vivo, *Adv. Mater.*, 29:1604894. DOI: 10.1002/adma.201770036. 52

[124] Lkhagvadulam, B., Kim, J. H., Yoon, I., and Shim, Y. K. 2013. Size-dependent photodynamic activity of gold nanoparticles conjugate of water soluble purpurin-18-N-methyl-d-glucamine, *Biomed. Res. Int.*, 720579. DOI: 10.1155/2013/720579. 52

[125] Oo, M. K., Yang, X., Du, H., and Wang, H. 2008. 5-aminolevulinic acid-conjugated gold nanoparticles for photodynamic therapy of cancer, *Nanomedicine (Lond)*, 3:777–786. DOI: 10.2217/17435889.3.6.777. 52

[126] Zhang, Z., Wang, S., Xu, H., Wang, B., and Yao, C. 2015. Role of 5-aminolevulinic acid-conjugated gold nanoparticles for photodynamic therapy of cancer, *J. Biomed, Opt.*, 20:51043. DOI: 10.1117/1.jbo.20.5.051043. 52, 53

[127] Meyers, J. D., Cheng, Y., Broome, A. M., Agnes, R. S., Schluchter, M. D., Margevicius, S., Wang, X., Kenny, M. E., Burda, C., and Basilion, J. P. 2015. Peptide-targeted gold nanoparticles for photodynamic therapy of brain cancer, *Part Part Syst. Charact.*, 32:448–457. DOI: 10.1002/ppsc.201400119. 52

[128] Nombona, N., Maduray, K., Antunes, E., Karsten, A., and Nyokong, T. 2012. Synthesis of phthalocyanine conjugates with gold nanoparticles and liposomes for photodynamic therapy, *J. Photochem. Photobiol. B*, 107:35–44. DOI: 10.1016/j.jphotobiol.2011.11.007. 52

[129] Shrestha, S., Wu, J., Sah, B., Vanasse, A., Cooper, L. N., Ma, L., Li, G., Zheng, H., Chen, W., and Antosh, M. P. 2019. X-ray induced photodynamic therapy with copper-cysteamine nanoparticles in mice tumors, *Proc. Natl. Acad. Sci. U. S. A.*, 116:16823–16828. DOI: 10.1073/pnas.1900502116. 52

[130] Bharathiraja, S., Moorthy, M. S., Manivasagan, P., Seo, H., Lee, K. D., and Oh, J. 2017. Chlorin e6 conjugated silica nanoparticles for targeted and effective photodynamic therapy, *Photodiagn. Photodyn. Ther.*, 19:212–220. DOI: 10.1016/j.pdpdt.2017.06.001. 53, 54

[131] Huang, P., Lin, J., Wang, S., Zhou, Z., Li, Z., Wang, Z., Zhang, C. Y. X., Niu, G., Yang, M., Cui, D., and Chen, X. 2013. Photosensitizer-conjugated silica-coated gold nanoclusters for fluorescence imaging-guided photodynamic therapy, *Biomaterials*, 34:4643–4654. DOI: 10.1016/j.biomaterials.2013.02.063. 53

[132] Brezaniova, I., Zaruba, K., Kralova, J., Sinica, A., Adamkova, H., Ulbrich, P., Pouckova, P., Hruby, M., Stepanek, P., and Kral, V. 2018. Silica-based nanoparticles are efficient delivery systems for temoporfin, *Photodiagnosis Photodyn. Ther.*, 21:275–284. DOI: 10.1016/j.pdpdt.2017.12.014. 53

[133] Gary-Bobo, M., Hocine, O., Brevet, D., Maynadier, M., Raehm, L., Richeter, S., Charasson, V., Loock, B., Morere, A., Maillard, P., Garcia, M., and Durand, J. O. 2012. Cancer therapy improvement with mesoporous silica nanoparticles combining targeting, drug delivery and PDT, *Int. J. Pharm.*, 423:509–515. DOI: 10.1016/j.ijpharm.2011.11.045. 53

[134] Kamkaew, A., Cheng, L., Goel, S., Valdovinos, H. F., Barnhart, T. E., Liu, Z., and Cai, W. 2016. Cerenkov radiation induced photodynamic therapy using chlorin e6-loaded hollow mesoporous silica nanoparticles, *ACS Appl. Mater. Interf.*, 8:26630-26637. DOI: 10.1021/acsami.6b10255. 53, 54

[135] Tu, J., Wang, T., Shi, W., Wu, G., Tian, X., Wang, Y., Ge, D., and Ren, L. 2012. Multifunctional ZnPc-loaded mesoporous silica nanoparticles for enhancement of photodynamic therapy efficacy by endolysosomal escape, *Biomaterials*, 33:7903–7914. DOI: 10.1016/j.biomaterials.2012.07.025. 53

[136] He, X., Wu, X., Wang, K., Shi, B., and Hai, L. 2009. Methylene blue-encapsulated phosphonate-terminated silica nanoparticles for simultaneous in vivo imaging and photodynamic therapy, *Biomaterials*, 30:5601–5609. DOI: 10.1016/j.biomaterials.2009.06.030. 53

[137] Brevet, D., Gary-Bobo, M., Raehm, L., Richeter, S., Hocine, O., Amro, K., Loock, B., Gouleaud, P., Frochot, C., Morere, A., Maillard, P., Garcia, M., and Durand, J.

O. 2009. Mannose-targeted mesoporous silica nanoparticles for photodynamic therapy, *Chem. Commun. (Camb)*, pages 1475–1477. DOI: 10.1039/b900427k. 53

[138] Hocine, O., Gary-Bobo, M., Brevet, D., Maynadier, M., Fontanel, S., Raehm, L., Richeter, S., Loock, B., Couleaud, P., Frochot, C., Charnay, C., Derrien, G., Smaihi, M., Sahmoune, A., Morere, A., Maillard, P., Garcia, M., and Durand, J. O. 2010. Silicalites and mesoporous silica nanoparticles for photodynamic therapy, *Int. J. Pharm.*, 402:221–230. DOI: 10.1016/j.ijpharm.2010.10.004. 53

[139] Benachour, H., Seve, A., Bastogne, T., Frochot, C., Vanderesse, R., Jasniewski, J., Miladi, I., Billotey, C., Tillement, O., Lux, F., and Heyob, M. B. 2012. Multifunctional peptide-conjugated hybrid silica nanoparticles for photodynamic therapy and MRI, *Theranostics*, 2:889–904. DOI: 10.7150/thno.4754. 53

[140] Qian, J., Wang, D., Cai, F., Zhan, Q., Wang, Y., and He, S. 2012. Photosensitizer encapsulated organically modified silica nanoparticles for direct two-photon photodynamic therapy and in vivo functional imaging, *Biomaterials*, 33:4851–4860. DOI: 10.1016/j.biomaterials.2012.02.053. 53, 55

[141] Kim, S., Ohulchanskyy, T. Y., Bharali, D., Chen, Y., Pandey, R. K., and Prasad, P. N. 2009. Organically modified silica nanoparticles with intraparticle heavy-atom effect on the encapsulated photosensitizer for enhanced efficacy of photodynamic therapy, *J. Phys. Chem. C Nanomater. Interf.*, 113:12641–12644. DOI: 10.1021/jp900573s. 53

[142] Kim, S., Ohulchanskyy, T. Y., Pudavar, H. E., Pandey, R. K., and Prasad, P. N. 2007. Organically modified silica nanoparticles co-encapsulating photosensitizing drug and aggregation-enhanced two-photon absorbing fluorescent dye aggregates for two-photon photodynamic therapy, *J. Am. Chem. Soc.*, 129:2669–2675. DOI: 10.1021/ja0680257. 53

[143] Ohulchanskyy, T. Y., Roy, I., Goswami, L. N., Chen, Y., Bergey, E. J., Pandey, R. K., Oseroff, A. R., and Prasad, P. N. 2007. Organically modified silica nanoparticles with covalently incorporated photosensitizer for photodynamic therapy of cancer, *Nano Lett.*, 7:2835–2842. DOI: 10.1021/nl0714637. 53

CHAPTER 5

3D in vitro Cancer Models

5.1 INTRODUCTION

Although numerous non-mammalian *in vivo* models including fruit fly, zebra fish, amphibians, as well as chicken embryos have demonstrated potential for use in PDT investigations, the emergence of different *in vitro* 3D cancer models has drawn enormous interest due to the ability of the systems to recapitulate important aspects of a solid tumor tissue such as tumor-stromal cell interactions while avoiding the use of living species [1–3]. The advantages that 3D models hold over the conventional 2D cell culture systems include enhancement in expression of differentiated functions, improved organization of cell/tissue, anti-apoptotic signaling, multicellular resistance, expression of hypoxic conditions, as well as restricted drug penetration. Moreover, it is possible to manipulate cellular and stromal characteristics of the 3D culture systems in order to produce a more representative model for the assessment of therapeutic response as a substitute for animal model testing [4–9].

Several methods can be used to develop 3D culture models. These include multicellular aggregates, culturing of cells on inserts, or embedding cells in an artificial nanofibrous matrix or scaffold generated from natural or synthetic material [10].

5.2 THREE-DIMENSIONAL (3D) IN VITRO CELL CULTURE MODELS

5.2.1 SPHEROIDS

Spheroids (also known as spheres, nodules, or micronodules) are well-rounded 3D cancer models with diameters that are normally several hundred μm long [11–13]. Such models are developed either through growing cells in low adhesion environments (e.g., plates, hanging drop methods, etc.) which allows them to adopt a spherical shape via aggregation [14–16] or by implanting the cells within a 3D matrix. Spheroid 3D models are quite comparable to *in vivo* tissues owing to the closely packed cellular arrangement within them which enables cell-cell contact in addition to a lower rate of drug and oxygen diffusion through them [17]. Besides therapeutic response, spheroids can also be employed for the investigation of various specific properties associated with 3D constructs, e.g., invasive characteristics, variations in the degree of cellular dependence on growth factors, improved luminal survival due to induction of anti-apoptotic and pro-proliferative signals, and the competency to avert growth arrest as a result of these pro-

proliferative signals [1, 18]. Spheroids have been involved in many photosensitizer PDT studies and also more recently in PCI assessments [19–27].

5.2.2   CELL DERIVED MATRICES (CDMS)

CDMs are often generated through culturing of ECM protein excreting cells on the surfaces of pre-coated scaffolds or as monolayers (2D)/multicellular aggregates (3D) in order to allow sufficient ECM deposition. Once this step is completed, the cellular components in the ECM are removed via decellularization processing. This minimizes the risk of experiencing adverse immunological responses [28]. Culturing cells on CDMs drives the 3D platform close to *in vivo* models since the cellular morphologies resemble those seen *in vivo* due to the formation of specified 3D matrix adhesions that are also found in the animal models [29].

5.2.3   MICROFLUIDIC DEVICES

The microfluidic technology also named Lab-on-a-chip (LOC) [30] is a profound system that enables 3D cell cultures as well as cell-based assays to be developed in complex microenvironments that are controllable, reproducible, and optimizable [31]. The key features associated with this type of technology are: (1) exhibition of micro-scale dimensions that are well compatible with the microstructures that exist in the microenvironments of *in vivo* cancer models; (2) cost-effectiveness, due to requirement of small sample quantities which in turn lead to low reagent, drug, and assay consumption; (3) high O_2 diffusion attributed to use of certain substrates with the device which ultimately influences cellular proliferation; and (4) accommodation for numerous applications, e.g., cell culture and sampling, fluid control, cell capturing, as well as cell lysis and detection in a single microfluidic device [32, 33]. Microfluidic devices are mainly named based on the types of substrates used to produce them. Examples of the different microfluidic devices that have been utilized for the development of 3D cell cultures include glass/silicon-based, polymer-based, and paper-based platforms [31, 34]. This technology has also been used in PDT studies to determine the efficiency of the treatment in the presence of photosensitizers, methylene blue, and 5-Aminolaevulinic acid (5-ALA), as well Gold nanoparticles (AuNPs) [19, 35].

5.2.4   SCAFFOLDS

3D scaffolds are composed of a nano-fibrous matrix that provides an environment which supports cancer cell proliferation, growth, and migration [36]. Compared to spheroids, scaffolds are able to imitate tumor heterogeneity and control of the 3D dimensions. Other advantages of scaffolds include the ability to regulate the degree of cellular migration, proliferation, and aggregation through manipulation of the scaffold's surface properties, configuration, and porosity [36, 37]. Such properties also make scaffolds suitable candidates for nanocarrier delivery investigations, as shown by Lopez et al. (2016) who placed a specific focus on the diffusion properties of liposomes and micelles in a 3D collagen scaffold model [38]. Other studies have also examined PLGA and lipid/liposomes as nanocarriers for photosensitizers such as enzophenothiazinium dye (EtNBS), Zinc phthalocyanine (ZnPc), and Indocyanine green in PDT studies, involving cancer spheroids in scaffold models [39–41]. Scaffolds can be categorized into natural or synthetic scaffolds depending on the materials incorporated into them [42].

5.2.5   NATURAL SCAFFOLDS

Natural scaffolds are mainly hydrogels that consist of mostly water [43] as well as natural components which include collagen type 1, Matrigel (a gelatinous complex protein mixture), agarose, elastin, laminin, and hyaluronic acid [18, 44]. Although the large volume of excess fluid within the hydrogels makes them mechanically weak, cellular movement and proliferation within such environment are made easy [36]. However, the low density of these models is not representative of the density observed in the environment that surrounds the tumor cells *in vivo* [36]. This problem can be partially resolved by the remodeling of the hydrogel to increase cell-matrix interaction and matrix density [45].

The introduction of the plastic compression technology into the 3D model preparation process has led to the fabrication of better biomimetic scaffolds with increased cell and collagen density due to the elimination of interstitial fluid from the hydrogel scaffold [46]. The collagen density in these compressed hydrogels (c. 10% wt/wt) is comparable to physiological values. The stiffness and density of collagen in these scaffolds not only affects cellular morphology and rate of proliferation [47], but also enables the hypoxic core that is present *in vivo* to be replicated *in vitro* as a result of reduced oxygen diffusion through the denser matrix [36]. Figure 5.1 demonstrates some of the different methods that could be used to develop *in vitro* 3D cancer models. Some cancer studies involving hydrogels have used compressed collagen hydrogels known as "tumoroid" models which consist of a central cancer mass surrounded by a multi-cellular stroma to investigate cancer progression as well as the uptake of nanoparticles and their roles in improving drug delivery [9, 48, 49]. Simple compressed 3D collagen models containing only cancer cells such as those of ovarian and renal cancer have too been employed in PCI studies and clinical trials (as personalized treatment screening tools) [27, 50, 51]. The construction of tumoroids was originally demonstrated by Nyga et al. (2013), who used plastic compression of collagen type I to create a colorectal tumoroid construct [52].

5.2.6   SYNTHETIC SCAFFOLDS

Synthetic scaffolds can be manufactured from polymers such as polyactide (PLA), polyglycolide (PGA), and co-polymers (PLGA) [42] which are biodegradable and have the capacity to be molded into various structures, e.g., mesh, fibers, and sponge [53]. In terms of mechanical structure, synthetic scaffolds are stronger compared to natural scaffolds and are able to almost exactly reproduce the biomolecular structures observed *in vivo* [54]. However, one disadvantage associated with these polymers is the weaker cell adhesions therefore surface modifications are needed to overcome this issue [36].

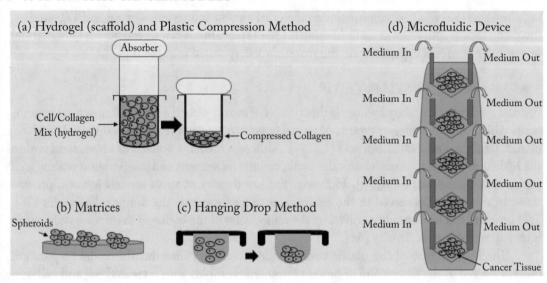

Figure 5.1: Examples of *in vitro* 3D cancer cultures developed using various techniques. (a) Cancer cells seeded in hydrogel scaffold which could then undergo plastic compression to form compressed 3D cancer models. (b) Formation of spheroids on matrix containing growth factors. (c) Spheroid formation using the hanging drop method where the base which cell suspension is applied on is flipped over allowing the suspension to hang from the base. The cells initially hang in single forms in suspension but gradually develop into spheroids. (d) Growth of cancer tissue within the chambers of a microfluidic device. The chambers have an inlet and outlet system allowing cell culture medium to be introduced via the inlet and removed via the outlet. All of the materials are usually housed in a polydimethylsiloxane (PDMS) membrane that is normally coated with ECM to support tissue growth.

5.3   REFERENCES

[1] Antoni, D., Burckel, H., Josset, E., and Noel, G. 2015. Three-dimensional cell culture: A breakthrough in vivo, *Int. J. Mol. Sci.*, 16:5517–5527. DOI: 10.3390/ijms16035517. 71, 72

[2] Chen, J., Wang, J., Zhang, Y., Chen, D., Yang, C., Kai, C., Wang, X., Shi, F., and Dou, J. 2014. Observation of ovarian cancer stem cell behavior and investigation of potential mechanisms of drug resistance in three-dimensional cell culture, *J. Biosci. Bioeng.*, 118:214–222. DOI: 10.1016/j.jbiosc.2014.01.008. 71

[3] Kucinska, M., Murias, M., and Nowak-Sliwinska, P. 2017. Beyond mouse cancer models: Three-dimensional human-relevant in vitro and non-mammalian in vivo models for

photodynamic therapy, *Mutat. Res.*, 773:242–262. DOI: 10.1016/j.mrrev.2016.09.002. 71

[4] van Duinen, V., Trietsch, S. J., Joore, J., Vulto, P., and Hankemeier, T. 2015. Microfluidic 3D cell culture: From tools to tissue models, *Curr. Opin. Biotechnol.*, 35:118–126. DOI: 10.1016/j.copbio.2015.05.002. 71

[5] Huh, D., Hamilton, G. A., and Ingber, D. E. 2011. From 3D cell culture to organs-on-chips, *Trends Cell Biol.*, 21:745–754. DOI: 10.1016/j.tcb.2011.09.005. 71

[6] Shin, C. S., Kwak, B., Han, B., and Park, K. 2013. Development of an in vitro 3D tumor model to study therapeutic efficiency of an anticancer drug, *Mol. Pharm.*, 10:2167–2175. DOI: 10.1021/mp300595a. 71

[7] Celli, J. P., Rizvi, I., Blanden, A. R., Massodi, I., Glidden, M. D., Pogue, B. W., and Hasan, T. 2014. An imaging-based platform for high-content, quantitative evaluation of therapeutic response in 3D tumour models, *Sci. Rep.*, 4:3751. DOI: 10.1038/srep03751. 71

[8] Gu, L. and Mooney, D. J. 2016. Biomaterials and emerging anticancer therapeutics: Engineering the microenvironment, *Nat. Rev. Cancer*, 16:56–66. DOI: 10.1038/nrc.2015.3. 71

[9] Ricketts, K., Cheema, U., Nyga, A., Castoldi, A., Guazzoni, C., Magdeldin, T., Emberton, M., Gibson, A., Royle, G., and Loizidou, M. 2014. A 3D in vitro cancer model as a platform for nanoparticle uptake and imaging investigations, *Small*, 10:3954–3961. DOI: 10.1002/smll.201400194. 71, 73

[10] Kimlin, L. C., Casagrande, G., and Virador, V. M. 2013. In vitro three-dimensional (3D) models in cancer research: An update, *Mol. Carcinog.*, 52:167–182. DOI: 10.1002/mc.21844. 71

[11] Weiswald, L. B., Bellet, D., and Dangles-Marie, V. 2015. Spherical cancer models in tumor biology, *Neoplasia*, 17:1–15. DOI: 10.1016/j.neo.2014.12.004. 71

[12] Kutys, M. L., Doyle, A. D., and Yamada, K. M. 2013. Regulation of cell adhesion and migration by cell-derived matrices, *Exp. Cell Res.*, 319:2434–2439. DOI: 10.1016/j.yexcr.2013.05.030. 71

[13] Evans, C. 2015. Three-dimensional in vitro cancer spheroid models for photodynamic therapy: Strengths and opportunities, *Front. Phys.*, 3:15. DOI: 10.3389/fphy.2015.00015. 71

[14] Charoen, K., Fallica, B., Colson, Y., Zaman, M., and Grinstaff, M. 2014. Embedded multicellular spheroids as a biomimetic 3D cancer model for evaluating drug and drug-device combinations, *Biomaterials*, 35:2264–2271. DOI: 10.1016/j.biomaterials.2013.11.038. 71

[15] Till, U., Gibot, L., Vicendo, P., Rols, M., Gaucher, M., Violleau, F., and Mingotaud, A. 2016. Crosslinked polymeric self-assemblies as an efficient strategy for photodynamic therapy on a 3D cell culture, *RSC Adv.*, 6:69984–69998. DOI: 10.1039/c6ra09013c. 71

[16] Hinger, D., Navarro, F., Käch, A., Thomann, J., Mittler, F., Couffin, A., and Maake, C. 2016. Photoinduced effects of m-tetrahydroxyphenylchlorin loaded lipid nanoemulsions on multicellular tumor spheroids, *J. Nanobiotechnol.*, 14:68. DOI: 10.1186/s12951-016-0221-x. 71

[17] Mehta, G., Hsiao, A., Ingram, M., Luker, G., and Takayama, S. 2012. Opportunities and challenges for use of tumor spheroids as models to test drug delivery and efficacy, *J. Control. Release*, 164:192–204. DOI: 10.1016/j.jconrel.2012.04.045. 71

[18] Herrmann, D., Conway, J. R., Vennin, C., Magenau, A., Hughes, W. E., Morton, J. P., and Timpson, P. 2014. Three-dimensional cancer models mimic cell-matrix interactions in the tumour microenvironment, *Carcinogenesis*, 35:1671–1679. DOI: 10.1093/carcin/bgu108. 72, 73

[19] Chen, Y. C., Lou, X., Zhang, Z., Ingram, P., and Yoon, E. 2015. High-throughput cancer cell sphere formation for characterizing the efficacy of photo dynamic therapy in 3D cell cultures, *Sci. Rep.*, 5:12175. DOI: 10.1038/srep12175. 72

[20] Rizvi, I., Celli, J. P., Evans, C. L., Abu-Yousif, A. O., Muzikansky, A., Pogue, B. W., Finkelstein, D., and Hasan, T. 2010. Synergistic enhancement of carboplatin efficacy with photodynamic therapy in a three-dimensional model for micrometastatic ovarian cancer, *Cancer Res.*, 70:9319–9328. DOI: 10.1158/0008-5472.can-10-1783. 72

[21] Evans, C. L., Abu-Yousif, A. O., Park, Y. J., Klein, O. J., Celli, J. P., Rizvi, I., Zheng, X., and Hasan, T. 2011. Killing hypoxic cell populations in a 3D tumor model with EtNBS-PDT, *PLoS One*, 6:e23434. DOI: 10.1371/journal.pone.0023434. 72

[22] Rowlands, C. J., Wu, J., Uzel, S. G., Klein, O., Evans, C. L., and So, P. T. 2014. 3D-resolved targeting of photodynamic therapy using temporal focusing, *Laser Phys. Lett.*, 11:115605. DOI: 10.1088/1612-2011/11/11/115605. 72

[23] Rizvi, I., Anbil, S., Alagic, N., Celli, J., Zheng, L. Z., Palanisami, A., Glidden, M. D., Pogue, B. W., and Hasan, T. 2013. PDT dose parameters impact tumoricidal durability and cell death pathways in a 3D ovarian cancer model, *Photochem. Photobiol.*, 89:942–952. DOI: 10.1111/php.12065. 72

[24] Huygens, A., Huyghe, D., Bormans, G., Verbruggen, A., Kamuhabwa, A. R., Roskams, T. and de Witte, P. A. M. 2003. Accumulation and photocytotoxicity of hypericin and analogs in two- and three-dimensional cultures of transitional cell carcinoma cells, *Photochem. Photobiol.*, 78:607–614. DOI: 10.1562/0031-8655(2003)0780607aapoha2.0.co2. 72

[25] Qiu, K., Wang, J., Song, C., Wang, L., Zhu, H., Huang, H., Huang, J., Wang, H., Ji, L., and Chao, H. 2017. Crossfire for two-photon photodynamic therapy with fluorinated ruthenium (II) photosensitizers ACS, *Appl. Mater. Interf.*, 9:18482–18492. DOI: 10.1021/acsami.7b02977. 72

[26] Anbil, S., Rizvi, I., Celli, J. P., Alagic, N., Pogue, B. W., and Hasan, T. 2013. Impact of treatment response metrics on photodynamic therapy planning and outcomes in a threedimensional model of ovarian cancer, *J. Biomed. Opt.*, 18:098004. DOI: 10.1117/1.jbo.18.9.098004. 72

[27] Mohammad Hadi, L., Yaghini, E., Stamati, K., Loizidou, M., and MacRobert, A. J. 2018. Therapeutic enhancement of a cytotoxic agent using photochemical internalisation in 3D compressed collagen constructs of ovarian cancer, *Acta Biomaterialia*, 81:80–92. DOI: 10.1016/j.actbio.2018.09.041. 72, 73

[28] Fitzpatrick, L. E. and McDevitt, T. C. 2015. Cell-derived matrices for tissue engineering and regenerative medicine applications, *Biomater. Sci.*, 3:12–24. DOI: 10.1039/c4bm00246f. 72

[29] Geraldo, S., Simon, A., Elkhatib, N., Louvard, D., Fetler, L., and Vignjevic, D. 2012. Do cancer cells have distinct adhesions in 3D collagen matrices and in vivo? *Eur. J. Cell Biol.*, 91:63–81. DOI: 10.1016/j.ejcb.2012.07.005. 72

[30] Ying, L. and Wang, Q. 2013. Microfluidic chip-based technologies: Emerging platforms for cancer diagnosis, *BMC Biotechnol.*, 13:76. DOI: 10.1186/1472-6750-13-76. 72

[31] Li, X. J., Valadez, A. V., Zuo, P., and Nie, Z. 2012. Microfluidic 3D cell culture: Potential application for tissue-based bioassays, *Bioanalysis*, 4:1509–1525. DOI: 10.4155/bio.12.133. 72

[32] Halldorsson, S., Lucumi, E., Gomez-Sjoberg, R., and Fleming, R. M. T. 2015. Advantages and challenges of microfluidic cell culture in polydimethylsiloxane devices, *Biosens. Bioelectron.*, 63:218–231. DOI: 10.1016/j.bios.2014.07.029. 72

[33] Gupta, N., Liu, J. R., Patel, B., Solomon, D. E., Vaidya, B., and Gupta, V. 2016. Microfluidicsbased 3D cell culture models: Utility in novel drug discovery and delivery research, *Bioeng. Trans. Med.*, 1:63–81. 72

[34] Tanyeri, M. and Tay, S. 2018. Viable cell culture in PDMS-based microfluidic devices, *Meth. Cell Biol.*, 148:3–33. DOI: 10.1016/bs.mcb.2018.09.007. 72

[35] Yang, Y., Yang, X., Zou, J., Jia, C., Hu, Y., Du, H., and Wang, H. 2015. Evaluation of photodynamic therapy efficiency using an in vitro three-dimensional microfluidic breast cancer tissue model, *Lab. Chip*, 15:735–744. DOI: 10.1039/c4lc01065e. 72

[36] Nyga, A., Cheema, U., and Loizidou, M. 2011. 3D tumour models: Novel in vitro approaches to cancer studies, *J. Cell Commun. Signal*, 5:239–248. DOI: 10.1007/s12079-011-0132-4. 72, 73

[37] Ng, R., Zhang, R., Yang, K. K., Liua, N., and Yang, S. T. 2012. Three-dimensional fibrous scaffolds with microstructures and nanotextures for tissue engineering, *RSC Adv.*, 2:10110–10124. DOI: 10.1039/c2ra21085a. 72

[38] López-Dávila, V., Magdeldin, T., Welch, H., Dwek, M. V., Uchegbu, I., and Loizidou, M. 2016. Efficacy of DOPE/DC-cholesterol liposomes and GCPQ micelles as AZD6244 nanocarriers in a 3D colorectal cancer in vitro model, *Nanomedicine*, 11:331–344. DOI: 10.2217/nnm.15.206. 72

[39] Hung, H. I., Klein, O. J., Peterson, S. W., Rokosh, S. R., Osseiran, S., Nowell, N. H., and Evans, C. L. 2016. PLGA nanoparticle encapsulation reduces toxicity while retaining the therapeutic efficacy of EtNBS-PDT in vitro, *Sci. Rep.*, 6:33234. DOI: 10.1038/srep33234. 73

[40] Lee, J., Kim, J., Jeong, M., Lee, H., Goh, U., Kim, H., Kim, B., and Park, J. H. 2015. Liposomebased engineering of cells to package hydrophobic compounds in membrane vesicles for tumor penetration, *Nano Lett.*, 15:2938–2944. DOI: 10.1021/nl5047494. 73

[41] Wang, Y., Xie, Y., Li, J., Peng, Z. H., Sheinin, Y., Zhou, J., and Oupicky, D. 2017. Tumor-penetrating nanoparticles for enhanced anticancer activity of combined photodynamic and hypoxia-activated therapy, *ACS Nano*, 11:2227–2238. DOI: 10.1021/acsnano.6b08731. 73

[42] Asti, A. and Gioglio, L. 2014. Natural and synthetic biodegradable polymers: Different scaffolds for cell expansion and tissue formation, *Int. J. Artif. Organs*, 37:187–205. DOI: 10.530/ijao.5000307. 73

[43] Geckil, H., Xu, F., Zhang, X., Moon, S., and Demirci, U. 2010. Engineering hydrogels as extracellular matrix mimics, *Nanomedicine*, 5:469–484. DOI: 10.2217/nnm.10.12. 73

[44] Zhu, J. and Marchant, R. E. 2011. Design properties of hydrogel tissue-engineering scaffolds, *Expert Rev. Med. Dev.*, 8:607–626. DOI: 10.1586/erd.11.27. 73

[45] Cheema, U., Yang, S. Y., Mudera, V., Goldspink, G. G., and Brown, R. A. 2003. 3-D in vitro model of early skeletal muscle development, *Cell Motil. Cytoskeleton.*, 54:226–236. 73

[46] Brown, R., Wiseman, M., Chuo, C., Cheema, U., and Nazhat, S. 2005. Ultrarapid engineering of biomimetic materials and tissues: Fabrication of nano- and microstructures by plastic compression advanced, *Funct. Mater.*, 15:1762–1770. DOI: 10.1002/adfm.200500042. 73

[47] Paszek, M. J., Zahir, N., Johnson, K. R., Lakins, J. N., Rozenberg, G. I., Gefen, A., Reinhart-King, C. A., Margulies, S. S., Dembo, M., Boettiger, D., Hammer, D. A., and Weaver, V. M. 2005. Tensional homeostasis and the malignant phenotype, *Cancer Cell*, 8:241–254. DOI: 10.1016/j.ccr.2005.08.010. 73

[48] López-Dávila, V., Magdeldin, T., Welch, H., Dwek, M. V., Uchegbu, I., and Loizidou, M. 2016. Efficacy of DOPE/DC-cholesterol liposomes and GCPQ micelles as AZD6244 nanocarriers in a 3D colorectal cancer in vitro model, *Nanomedicine (Lond.)*, 11:331–344. DOI: 10.2217/nnm.15.206. 73

[49] Pape, J., Magdeldin, T., Ali, M., Walsh, C., Lythgoe, M., Emberton, M., and Cheema, U. 2019. Cancer invasion regulates vascular complexity in a three-dimensional biomimetic model, *Eur. J. Cancer*, 119:179–193. DOI: 10.1016/j.ejca.2019.07.005. 73

[50] Mohammad Hadi, L., Yaghini, E., MacRobert, A. J., and Loizidou, M. 2020. Synergy between photodynamic therapy and dactinomycin chemotherapy in 2D and 3D ovarian cancer cell cultures, *Int. J. Mol. Sci.*, 21:E3203. DOI: 10.3390/ijms21093203. 73

[51] Tran, M. G. B., Neves, J. B., Stamati, K., Redondo, P., Cope, A., Brew-Graves, C., Williams, N. R., Grierson, J., Cheema, U., Loizidou, M., and Emberton, M. 2019. Acceptability and feasibility study of patient-specific "tumouroids" as personalised treatment screening tools: Protocol for prospective tissue and data collection of participants with confirmed or suspected renal cell carcinoma, *Int. J. Surg. Protoc.*, 14:24–29. DOI: 10.1016/j.isjp.2019.03.019. 73

[52] Nyga, A., Loizidou, M., Emberton, M., and Cheema, U. 2013. A novel tissue engineered three-dimensional in vitro colorectal cancer model, *Acta Biomater.*, 9:7917–7926. DOI: 10.1016/j.actbio.2013.04.028. 73

[53] Dhandayuthapani, B., Yoshida, Y., Maekawa, T., and Kumar, D. S. 2011. Polymeric scaffolds in tissue engineering application: A review, *Int. J. Polym. Sci.*, pages 1–19. DOI: 10.1155/2011/290602. 73

[54] Perez-Castillejos, R. 2010. Replication of the 3D architecture of tissues, *Mater. Today*, 13:32–41. DOI: 10.1016/s1369-7021(10)70015-8. 73

CHAPTER 6

The Applications of PDT and PCI in 3D in vitro Cancer Models

6.1 INTRODUCTION

The use of 3D models to assess cellular response to PDT and PCI has become rather popular in recent years due to the capability of the models to overcome some of the limitations associated with the 2D cultures and enable rapid screening which could potentially reduce the requirement for costly *in vivo* experimentation. 3D cancer models, most commonly spheroids, have contributed majorly to photodynamic related research allowing various aspects such as uptake and therapeutic efficacy of a range of photosensitizers and anti-cancer drugs (for PDT and PCI) particularly at lower oxygen levels to be tested. This makes the model and the investigations more physiologically relevant as PDT efficacy relies on adequate oxygen supply which is lower in solid tumors [1]. Furthermore, the 3D tissue models could mimic the oxygen diffusion and consumption observed *in vivo* during PDT since the PDT process consumes molecular oxygen as molecular substrates undergo oxidisation [2].

6.2 PDT IN 3D CANCER MODELS

Spheroids also referred to as nodules and spheres have been employed widely for PDT studies mostly in the presence of natural scaffolds. Microfluidic devices have also made their entry into this field of research in recent years as advanced 3D models that are able to provide a dynamic environment for *in vitro* tumor growth and therapeutic efficacy investigations. Numerous studies have compared differences in the response of cells to PDT in 2D and 3D cultures.

Evans et al. (2011) used the ovarian cancer cell line OVCAR5 to develop Nodules on a Matrigel to study PDT effect using photosensitizer 5-ethylamino-9-diethylaminobenzo[a] phenothiazinium chloride (EtNBS) on the hypoxic cell populations in 3D models of ovarian cancer. The nodules underwent incubation with 500 nM EtNBS for 4.5 hr in order to allow the concentration of the photosensitizer in the core cell populations within the nodules. Exposure of the nodules to a light dose of 5 J/cm², enabled selective destruction of the nodules' core cells by EtNBS. Therefore, EtNBS was able to penetrate into the core of the nodules and destroy the usually difficult to treat hypoxic cell populations. Increasing the light doses, resulted in cell killing

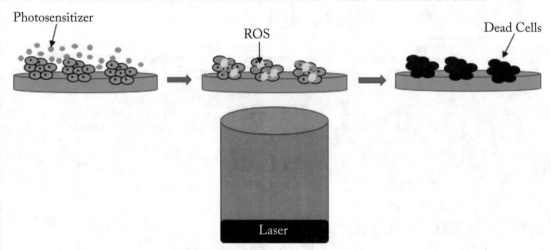

Figure 6.1: Application of PDT in cancer spheroids (also called spheres, nodules, micronodules) that have been developed on Matrigel coating. The spheroids are incubated with the photosensitizer for an adequate period before being irradiated using a laser of an appropriate wavelength which results in the production of reactive oxygen species (ROS) and ultimately cell death.

across the whole nodule, demonstrating that EtNBS-PDT is effective against both hypoxic and normoxic regions of a tumor. Another interesting finding of this study was the increase in uptake and cytotoxicity of EtNBS with the expansion of nodule size which was mainly attributed to the rise in hypoxic populations in the larger spheroids [3].

Numerous studies have focused on the effect of the Benzoporphyrin Derivative (BPD)-PDT on 3D micronodule and nodule models of ovarian cancer on matrigels using OVCAR5 cells [4–8]. For example, Rizvi et al. (2013) treated OVCAR5 micronodules that had been developed on growth factor reduced (GFR) Matrigel with 0.25 μM, 1 μM, and 10 μM BPD for 90-min before exposing them to irradiation with 690-nm fiber-coupled diode laser. The treatment caused significant reduction in cell viability in micronodules that underwent exposure to 0.25 μM BPD-PDT particularly 72 and 96 hr post treatment in comparison to micronodules that received 1 and 10 μM BPD-PDT. Surprisingly, models that were treated with the highest photosensitizer concentration (10 μM BPD-PDT) exhibited the poorest response, which may have resulted from aggregation of the photosensitizer, thus impairing the photosensitizing efficacy of BPD [6]. Figure 6.1 demonstrates the process of treating spheroids or nodules that have been grown on Matrigel.

In another study by Rizvi et al., BPD-PDT was combined with low-dose carboplatin (chemotherapeutic agent) to enhance treatment of OVCAR5 micronodules by carboplatin. The models were treated with carboplatin (40 mg/m^2) either prior to or after 1.25 μM BPD-PDT with 690-nm fiber-coupled diode laser (fluence of 5.0 J/cm^2). Treating the micronodules with

BPD-PDT prior to carboplatin led to a significant synergistic reduction in residual tumor volume and viability in comparison to PDT or carboplatin monotherapies and controls. The mean fraction viability achieved post combination of PDT and carboplatin was 0.45 vs. PDT (0.80) or carboplatin (0.92) alone. However, no synergistic effect was observed between the monotherapies and reverse treatment order (carboplatin treatment followed by BPD-PDT). The inability of carboplatin monotherapy to cause eradication of the micronodules led to the survival of cores that lacked sensitivity and permeability to chemotherapy. Furthermore, the study found that the 2D cultures significantly overestimated the OVCAR5 cells sensitivity to the treatments, which made the 3D models used more reliable tools for testing therapeutic efficacy [4]. Celli et al. (2014), on the other hand, discovered that BPD-PDT disaggregated the larger micronodules into numerous smaller nodules [8].

Spheroids of other types of human cancer cells such as epidermoid, breast, bladder, and cervical have also been developed in different supporting systems including microfluidic device, basement membrane coating, hydrogel, matrigel, agarose, and liquid overlay to test the effectiveness of various sensitizers other than BPD in treating the cancer cells through PDT [9–16].

Chen et al. (2015) used the microfluidic platform for the development of T47D breast cancer cell spheres on a chip. The spheres formed in individual microwells as a result of aggregation. The models were treated with photosensitizer methylene blue (10 μM) for 1 hr before being illuminated for different periods. Exposure of the spheres to light dose of 7.3 J/cm^2 for 10-min led to nearly 50% cell kill in the monolayer cultures while most of the cells in 3D spheres remained viable. Interestingly, a portion of the cells located in the center of the spheres were still viable even after treatment with the highest illumination period (1 hr) (44 Jcm^2). The larger spheres showed less sensitivity toward PDT than small spheres which indicated the impact of sphere size on the efficacy of PDT [9].

Another approach employed to support breast cancer spheroid development from SUM149 cells is the use of reconstituted basement membrane coated glass as shown by Aggarwal et al. (2015). This group treated the spheroids with BPD (1.5 μM) and/or mono-l-aspartyl chlorin e6 (NPe6) (40 μM) for 1 hr before irradiation with different doses of light at 690-nm (to induce BPD-PDT effects) and/or at 660-nm (to induce NPe6-PDT effects). The combination of the two sensitizers with BPD being activated first led to 19% cell viability using the highest light dose for both wavelengths (45 mJ/cm^2) while the same light dose resulted in 6% viability when NPe6 was activated first. Furthermore, the combination of BPD and NPe6 produced a greater response than the application of either sensitizer alone. Thus, the results indicated that targeting the lysosomes via NPe6 activation before the mitochondria via BPD activation enhanced the photokilling effects in the 3D models [11].

Wright et al. (2009), however, used unaggregated cells embedded in 3D hydrogel scaffold to study the effect of meta-tetrahydroxyphenyl chlorin-(mTHPC) mediated PDT on rat neurons and satellite glia compared to human adenocarcinoma cells (MCF-7). After incubating the cultures with mTHPC for 4 hr, the models underwent irradiation with white light (1 Jcm^2)

for 10-min. Overall, the neurons showed more resistance toward mTHPC-PDT than satellite glial or tumor cells with the 4 μg/mL concentration of mTHPC causing 48%, 39%, and 11% cell death in MCF-7, satellite glia, and neurons, respectively, after irradiation [12].

A different study investigated the effect of 5-Aminolevulinic acid (ALA)-PDT on nodules of A431 human epidermal carcinoma grown on matrigel. Following 4-hr incubation with ALA (1 mM), the nodules were irradiated with 635-nm light. Increasing the light doses from 5–80 J/cm^2 caused increasing cytotoxicity compared to controls with the higher light doses (40–80 J/cm^2) resulting in nearly complete cytotoxic destruction of the nodules. This group also used a smartphone camera to detect PpIX fluorescence. Remarkably, the device was able to distinguish the ALA treated nodules from the non-treated nodules through identifying the fluorescence signals from the PpIX [10].

When Huygens et al. (2007) tested PDT with hypericin and its iodinated derivatives (mono-iodohypericin or di-iodohypericin) at concentrations of 125 nM, 1 μM, or 10 μM (2 or 24-hr incubation) and light dose of 1.8 J/cm^2 (for 3 or 30-min) in 2D and 3D spheroid (on agarose coating) cultures of RT112 bladder carcinoma cells, it was discovered that prolonging the irradiation period and increasing the photosensitizer concentration led to phototoxicity enhancement. However, hypericin and mono-iodohypericin were more effective in inducing antiproliferative effect post PDT compared to di-iodohypericin in both spheroids and monolayer cultures with the effect being stronger in the monolayer cultures. Overall, the photocytotoxic effect of all three sensitizers was dramatically lower in the spheroids. Although di-iodohypericin concentrated to a greater degree in the monolayer and spheroid cultures than hypericin and mono-iodohypericin, the accumulation was mainly confined to the outermost section of the spheroids [13].

Several studies have investigated ruthenium complexes as anticancer agents in two-photon PDT of 3D human cervical carcinoma spheroid models developed from HeLa cells [14–16]. For example, Liu et al. (2015) employed four ruthenium (II) polypyridyl complexes (RuL1–RuL4) as mitochondria-targeted two photon photodynamic anticancer agents and tested them in HeLa spheroid cultures on agarose coating. Subjecting the spheroids to treatment with 10 μM RuL1–RuL4 for 3 days followed by two photon irradiation resulted in effective growth inhibition as well as gradual decrease in spheroid volume with RuL4 showing the strongest inhibiting capability out of the complexes tested [15].

6.3 APPLICATION OF PDT INVOLVING NANOPARTICLES IN 3D CANCER MODELS

Nanoparticle (NP) delivery systems have been used extensively for improving tumor selectivity and bioavailability of sensitizers. Numerous studies have also tested such systems in 3D cancer models [17].

The regulation of oxygen and NP delivery to cells in 3D cultures via diffusion through the matrix makes these models suitable for studying nanoparticle efficacy for PDT. One important advantage of using the 3D models for PDT investigations involving NPs is their ability to display cell penetration issues experienced by NPs that are normally absent in 2D monolayer cultures [18]. The various NPs used for PDT studies in 3D models include inorganic (gold) (AuNPs), micelles, polymeric carriers, liposomes, and single-walled carbon nanotubes [19–28].

Yang et al. (2015) studied the synergistic enhancement of PDT with 5-ALA and AuNPs in 3D breast cancer tissue model consisting of human breast cancer cells (MCF-7) in addition to primary adipose derived stromal cells (ASCs) developed on a microfluidic device. Other aspects investigated included the agents' distribution profiles and impact of light penetration on PDT efficacy based on depth of the cancer tissue. The 3D cancer tissue models and monolayer MCF-7 cultures were incubated with 5-ALA (1 mM) dissolved in a serum free medium either with or without AuNPs for 4 hr. PDT with 5-ALA only resulted in 50% cell kill in the monolayer cultures and 17% cell kill in the 3D cultures in comparison to the 70% (2D culture) and 50% (3D culture) cell kill attained in the presence of AuNPs. Prolonging the illumination period led to destruction of most of the cells in the 2D culture both in the presence and absence of AuNPs while in the 3D cultures, the application of 5-ALA alone proved less effective causing 50% cell kill compared to the 90% kill achieved when 5-ALA was accompanied by AuNPs. Overall, a considerably higher cell death rate was observed in the monolayer culture than the 3D cancer tissue model across all illumination periods. Another interesting finding by this group was the homogenous response generated whereby the dead cells produced from the combined treatment with 5-ALA and AuNPs were distributed across the full thickness of the 3D cancer model, however following treatment with 5-ALA only PDT, most of the dead cell population were found within the superficial regions of the cancer tissue [19].

Till et al. (2016) developed 3D spheroid models of human colorectal carcinoma cell line (HCT-116) and human squamous cell carcinoma cell line (FaDu) using ultra-low attachment plates to test PDT efficacy with free and encapsulated forms of pheophorbide (Pheo) photosensitizer. With regards to the photosensitizer loaded NPs, Pheo was either encapsulated inside crosslinked polymeric micellar self-assemblies or un-crosslinked micellar NPs. The spheroids underwent incubation with photosensitizer in free on encapsulated form ([polymer] = 100 μM, [Pheo] = 3.33 μM) for 30-min prior to being exposed to irradiation with > 400-nm light (8.2 J/cm^2) for 8-min. Following the first cycle of irradiation, the spheroids underwent treatment with another 2 cycles of 8-min irradiation, once every 24 hr for 2 days. The free form of photosensitizer produced low PDT efficiency as a result of its aggregation in water thus, an inefficient response was achieved in FaDu spheroids and only a 35% reduction in the size of HCT-116 spheroids was observed following the third cycle of irradiation. The encapsulated forms of Pheo proved more effective in treating the spheroids using PDT than the free photosensitizer. Furthermore, the cross-linked systems were more efficient than the un-crosslinked nanovectors in terms of treatment capability via PDT. In contrast to the 3D model results, 2D cultures demonstrated better cell killing with the application of uncrosslinked micelles. According to the authors *in vivo* studies comparing cross-linked and un-crosslinked forms of chemotherapeutic drug carrying micelles found that crosslinked micelles were more effective than their un-crosslinked counterparts which is supportive of the outcome obtained in this study, particularly since 3D models replicate the *in vivo* response better than monolayer cultures [20].

Hung et al. (2016) endeavored to minimize the dark toxicity associated with EtNBS and assess its efficiency as a PDT sensitizer in monolayer and spheroid cultures of OVCAR5 cells by encapsulating the EtNBS in PLGA NPs. A reduction in dark toxicity was witnessed in monolayer cultures treated with EtNBS loaded nanoparticles compared to those treated with free EtNBS. According to the uptake studies conducted in the spheroids, PLGA-EtNBS diffused throughout the spheroids in a similar manner to free EtNBS. This study also demonstrated the release of EtNBS from PLGA after illumination with 635-nm laser light. Impressively, PDT using PLGA-EtNBS delivery displayed effectiveness even in the hypoxic cellular microenvironments in the spheroids, with the efficacy being comparable to that of free EtNBS [22].

Zhang et al. (2015) studied the induction of photothermal and two-photon PDT (PTT/TPPDT) effects in spheroid models of cervical cancer (HeLa cells) using Ru (II) complex loaded SWCNTs (Ru@SWCNTs). The spheroid models are particularly useful for such study since the closely packed cells within the spheroids allow the optical sectioning potentials of multiphoton PDT to be displayed. Incubating the spheroids with Ru@SWCNTs (50 μg/mL) and afterward exciting them with 808-nm (0.25 W/cm^2) laser resulted in a decrease in the mean diameter of the spheroids as well as a significant reduction in cell viability [23]. Liu et al. (2017), on the other hand, investigated PTT and PDT efficacy in monolayer and spheroid cultures of human brain carcinoma (U87 cells) by using chlorine e6 (Ce6) loaded reduced graphene oxide (rGO) nanocarriers. While both PTT and PDT showed effectiveness in the 2D cultures, only PTT exhibited a substantial treatment efficacy in the spheroid cultures [24].

Liposomes have also been taken advantage of in several PDT studies involving cancer spheroid models, due to their lipophilic/hydrophilic compound encapsulating capabilities as well as other benefits. In one study by Gaio et al. (2016), for example, photo-induced damage caused by two liposomal formulations of m-THPC, Foslip®, and Fospeg® in spheroid models of cervical cancer (HeLa cells) was compared to Foscan®. The results from the confocal fluorescence microscopy indicated that m-THPC mainly penetrated into the external cell layers of spheroids with Foslip® and Fospeg® showing slightly higher accumulation than Foscan®. Furthermore, a considerable reduction in cellular viability was detected in spheroids that underwent incubation with Foscan (8 μM) in the dark while spheroids that were incubated with the liposomal formulations did not demonstrate dark toxicity. After illumination with red light (22.5 J/cm^2), the cellular viabilities were found to be reduced significantly for all three formulations with Foslip® displaying the highest reduction in viability at each time point [25].

6.4   PCI IN 3D CANCER MODELS

O'Rouke et al. (2017) compared the sensitivity of neurons and glial cells to PCI in the presence of TPCS$_{2a}$/TPPS$_{2a}$ (photosensitizers) and bleomycin (chemotherapeutic drug) and 420-nm light (for 5-min) to that of PCI30 cells (head and neck squamous cell carcinoma) in 3D hydrogel models. According to the results, the neurons were more resistant to the PCI treatment than PCI30 cells and could withstand conditions that were otherwise sufficient for destroying the cancer cells, such as those induced by PCI. The degree of cell death displayed by the satellite glia on the other hand was similar to that shown by PCI30 cells [29].

Mohammad-Hadi et al. (2018) made use of the novel plastic compression technology to produce 3D compressed collagen ovarian cancer models consisting of SKOV3 or HEY cells (human ovarian cancer cells) to test PDT and PCI using TPPS$_{2a}$ and saporin (chemotherapeutic agent) as well as illumination with 420-nm blue lamp for different periods (3, 5, and 7-min). The testing was carried out on the models at two different stages, prior to formation of spheroids and following formation of spheroids. Although PCI was effective in both non-spheroid and spheroid cultures and resulted in a higher level of cell death compared to PDT or application of saporin alone, the reductions in cellular viability were lower for all conditions in the spheroid models in comparison to the non-spheroid models. Furthermore, the concentrations of TPPS$_{2a}$ and saporin required to attain significant treatment efficacy in the spheroid models were higher at 0.7 μg/mL and 40 nM (SKOV3 cells) and 0.5 μg/mL and 20 nM (HEY) cells compared to their non-spheroid counterparts 0.3 μg/mL and 20 nM (SKOV3 cells) and 0.4 μg/mL and 20 nM (HEY cells). PCI treatment reduced cell viabilities to 21% (SKOV3 cells) and 23% (HEY cells) in non-spheroid models and to 39% (SKOV3 cells) and 24% (HEY cells) in spheroid models after 5-min of illumination. The lower efficacy of PCI in the spheroid models compared to the non-spheroid models was attributed to limited drug penetration through the aggregates and the lower levels of oxygen in the spheroid models which adversely affected the photodynamic process involved in PCI. Interestingly, HEY spheroid models demonstrated more sensitivity toward PCI treatment for all illumination periods than the SKOV3 spheroid models despite showing the tendency to form larger aggregates than SKOV3 cells. Another remarkable finding was that in both non-spheroid and spheroid models, higher degree of apoptosis could be observed in contrast to the PDT treatment group which showed greater degree of necrosis in the cells. Moreover, the levels of necrosis and apoptosis induced in the PCI treated group were higher in the models that underwent incubation for 48 hr than those incubated for 24 hr following illumination [30].

In a subsequent study, Mohammad-Hadi et al. (2020) treated SKOV3 and HEY cell non-spheroid models with the highly toxic anti-cancer drug "dactinomycin" through PCI using the same photosensitizer (TPPS$_{2a}$). The use of PCI allowed effective treatment of the cancer cells with a very low concentration of dactinomycin (1 nM). The percentage viabilities of the cells were reduced to below 10% following 7-min of light illumination [31].

Martinez et al. (2017) assessed PCI effect with TPCS$_{2a}$/TPPS$_{2a}$ and saporin in 3D non-spheroid uncompressed collagen hydrogel culture of PC3 and MatLyLu prostate carcinoma cells. The cancer cells embedded in these scaffolds were treated prior to aggregating and forming spheroids. The Live-Dead assay results of this study demonstrated that PCI led to morphological changes in the cells, causing them to assume a more rounded shape post treatment. Furthermore, upon uptake by the monolayer cultures of the cells, $TPCS_{2a}/TPPS_{2a}$ and Saporin formed discrete granules which were consistent with endolysosomal localization. One-minute illumination with 405-nm laser led to dispersal of saporin [32].

6.5   REFERENCES

[1] Evans, C. 2015. Three-dimensional in vitro cancer spheroid models for photodynamic therapy: Strengths and opportunities, *Front. Phys.*, 3:15. DOI: 10.3389/fphy.2015.00015. 81

[2] Alemany-Ribes, M., Garcia-Diaz, M., Busom, M., Nonell, S., and Semino, C. E. 2013. Toward a 3D cellular model for studying in vitro the outcome of photodynamic treatments: Accounting for the effects of tissue complexity, *Tissue Eng. Part A*, 19:1665–1674. DOI: 10.1089/ten.tea.2012.0661. 81

[3] Evans, C. L., Abu-Yousif, A. O., Park, Y. J., Klein, O. J., Celli, J. P., Rizvi, I., Zheng, X., and Hasan, T. 2011. Killing hypoxic cell populations in a 3D tumor model with EtNBS-PDT, *PLoS One*, 6:e23434. DOI: 10.1371/journal.pone.0023434. 82

[4] Rizvi, I., Celli, J. P., Evans, C. L., Abu-Yousif, A. O., Muzikansky, A., Pogue, B. W., Finkelstein, D., and Hasan, T. 2010. Synergistic enhancement of carboplatin efficacy with photodynamic therapy in a three-dimensional model for micrometastatic ovarian cancer, *Cancer Res.*, 70:9319–9328. DOI: 10.1158/0008-5472.can-10-1783. 82, 83

[5] Rowlands, C. J., Wu, J., Uzel, S. G. M., Klein, O., Evans, C. L., and So, P. T. C. 2014. 3D-resolved targeting of photodynamic therapy using temporal focusing, *Laser Phys. Lett.*, 11:115605. DOI: 10.1088/1612-2011/11/11/115605. 82

[6] Rizvi, I., Anbil, S., Alagic, N., Celli, J., Zheng, L. Z., Palanisami, A., Glidden, M. D., Pogue, B. W., and Hasan, T. 2013. PDT dose parameters impact tumoricidal durability and cell death pathways in a 3D ovarian cancer model, *Photochem. Photobiol.*, 89:942–952. DOI: 10.1111/php.12065. 82

[7] Anbil, S., Rizvi, I., Celli, J. P., Alagic, N., Pogue, B. W., and Hasan, T. 2013. Impact of treatment response metrics on photodynamic therapy planning and outcomes in a three-dimensional model of ovarian cancer, *J. Biomed. Opt.*, 18:098004. DOI: 10.1117/1.jbo.18.9.098004. 82

[8] Celli, J. P., Rizvi, I., Blanden, A. R., Massodi, I., Glidden, M. D., Pogue, B. W., and Hasan, T. 2014. An imaging-based platform for high-content, quantitative evaluation of therapeutic response in 3D tumour models, *Sci. Rep.*, 4:3751. DOI: 10.1038/srep03751. 82, 83

[9] Chen, Y. C., Lou, X., Zhang, Z., Ingram, P., and Yoon, E. 2015. High-throughput cancer cell sphere formation for characterizing the efficacy of photo dynamic therapy in 3D cell cultures, *Sci. Rep.*, 5:12175. DOI: 10.1038/srep12175. 83

[10] Hempstead, J., Jones, D. P., Ziouche, A., Cramer, G. M., Rizvi, I., Arnason, S., Hasan, T., and Celli, J. P. 2015. Low-cost photodynamic therapy devices for global health settings: Characterization of battery-powered LED performance and smartphone imaging in 3D tumor models, *Sci. Rep.*, 5:10093. DOI: 10.1038/srep10093. 83, 84

[11] Aggarwal, N., Santiago, A. M., Kessel, D., and Sloane, B. F. 2015. Photodynamic therapy as an effective therapeutic approach in MAME models of inflammatory breast cancer, *Breast Cancer Res. Treat.*, 154:251–262. DOI: 10.1007/s10549-015-3618-6. 83

[12] Wright, K. E., Liniker, E., Loizidou, M., Moore, C., MacRobert, A. J., and Phillips, J. B. 2009. Peripheral neural cell sensitivity to mTHPC-mediated photodynamic therapy in a 3D in vitro model, *Br. J. Cancer*, 101:658–665. DOI: 10.1562/0031-8655(2003)0780607aapoha2.0.co2. 83, 84

[13] Huygens, A., Huyghe, D., Bormans, G., Verbruggen, A., Kamuhabwa, A. R., Roskams, T., and de Witte, P. A. M. 2003. Accumulation and photocytotoxicity of hypericin and analogs in two- and three-dimensional cultures of transitional cell carcinoma cells, *Photochem. Photobiol.*, 78:607–614. 83, 84

[14] Huang, H., Yu, B., Zhang, P., Huang, J., Chen, Y., Gasser, G., Ji, L., and Chao, H. 2015. Highly charged ruthenium(II) polypyridyl complexes as lysosome-localized photosensitizers for two-photon photodynamic therapy, *Angew Chem. Int. Ed.*, 54:14049–14052. 83, 84

[15] Liu, J., Chen, Y., Li, G., Zhang, P., Jin, C., Zeng, L., Ji, L., and Chao, H. 2015. Ruthenium(II) polypyridyl complexes as mitochondria-targeted two-photon photodynamic anticancer agents, *Biomaterials*, 56:140–153. DOI: 10.1016/j.biomaterials.2015.04.002. 83, 84

[16] Qiu, K., Wang, J., Song, C., Wang, L., Zhu, H., Huang, H., Huang, J., Wang, H., Ji, L., and Chao, H. 2017. Crossfire for two-photon photodynamic therapy with fluorinated ruthenium (II) photosensitizers, *ACS Appl. Mater. Interf.*, 9:18482–18492. DOI: 10.1021/acsami.7b02977. 83, 84

[17] Mohammad-Hadi, L., MacRobert, A. J., Loizidou, M., and Yaghini, E. 2018. Photodynamic therapy in 3D cancer models and the utilisation of nanodelivery systems, *Nanoscale*, 10:1570–1581. DOI: 10.1039/c7nr07739d. 84

[18] Zhao, J., Lu, H., Wong, S., Lu, M., Xiao, P., and Stenzel, M. H. 2017. Influence of nanoparticle shapes on cellular uptake of paclitaxel loaded nanoparticles in 2D and 3D cancer models. *Polym. Chem.*, 8:3317–26. DOI: 10.1039/c7py00385d. 85

[19] Yang, Y., Yang, X., Zou, J., Jia, C., Hu, Y., Du, H., and Wang, H. 2015. Evaluation of photodynamic therapy efficiency using an in vitro three-dimensional microfluidic breast cancer tissue model, *Lab. Chip*, 15:735–744. DOI: 10.1039/c4lc01065e. 85

[20] Till, U., Gibot, L., Vicendo, P., Rols, M. P., Gaucher, M., Violleau, F., and Mingotaud, A. F. 2016. Crosslinked polymeric self-assemblies as an efficient strategy for photodynamic therapy on a 3D cell culture, *RSC Adv.*, 6:69984–69998. DOI: 10.1039/c6ra09013c. 85, 86

[21] Hinger, D., Navarro, F., Kach, A., Thomann, J. S., Mittler, F., Couffin, A. C., and Maake, C. 2016. Photoinduced effects of m-tetrahydroxyphenylchlorin loaded lipid nanoemulsions on multicellular tumor spheroids, *J. Nanobiotechnol.*, 14:68. DOI: 10.1186/s12951-016-0221-x. 85

[22] Hung, H. I., Klein, O. J., Peterson, S. W., Rokosh, S. R., Osseiran, S., Nowell, N. H., and Evans, C. L. 2016. PLGA nanoparticle encapsulation reduces toxicity while retaining the therapeutic efficacy of EtNBS-PDT in vitro, *Sci. Rep.*, 6:33234. DOI: 10.1038/srep33234. 85, 86

[23] Zhang, P., Huang, H., Huang, J., Chen, H., Wang, J., Qiu, K., Zhao, D., Ji, L., and Chao, H. 2015. Noncovalent ruthenium(II) complexes-single-walled carbon nanotube composites for bimodal photothermal and photodynamic therapy with near-infrared irradiation, *ACS Appl. Mater. Interf.*, 7:23278–23290. DOI: 10.1021/acsami.5b07510. 85, 86

[24] Liu, J., Liu, K., Feng, L., Liu, Z., and Xu, L. 2017. Comparison of nanomedicine-based chemotherapy, photodynamic therapy and photothermal therapy using reduced graphene oxide for the model system, *Biomater. Sci.*, 5:331–340. DOI: 10.1039/c6bm00526h. 85, 86

[25] Gaio, E., Scheglmann, D., Reddi, E., and Moret, F. 2016. Uptake and photo-toxicity of Foscan®, Foslip®and Fospeg®in multicellular tumor spheroids, *J. Photochem. Photobiol. B*, 161:244–252. DOI: 10.1016/j.jphotobiol.2016.05.011. 85, 86

[26] Lee, J., Kim, J., Jeong, M., Lee, H., Goh, U., Kim, H., Kim, B., and Park, J. H. 2015. Liposomebased engineering of cells to package hydrophobic compounds in membrane vesicles for tumor penetration, *Nano Lett.*, 15:2938–2944. DOI: 10.1021/nl5047494. 85

[27] Xiao, Z., Hansen, C. B., Allen, T. M., Miller, G. G., and Moore, R. B. 2005. Distribution of photosensitizers in bladder cancer spheroids: Implications for intravesical instillation

of photosensitizers for photodynamic therapy of bladder cancer, *J. Pharm. Pharm. Sci.*, 8:536–543. 85

[28] Wang, Y., Xie, Y., Li, J., Peng, Z. H., Sheinin, Y., Zhou, J., and Oupicky, D. 2017. Tumor-penetrating nanoparticles for enhanced anticancer activity of combined photo-dynamic and hypoxia-activated therapy, *ACS Nano*, 11:2227–2238. DOI: 10.1021/ac-snano.6b08731. 85

[29] O'Rourke, C., Hopper, C., MacRobert, A. J., Phillips, J. B., and Woodhams, J. H. 2017. Could clinical photochemical internalisation be optimised to avoid neuronal toxicity? *Int. J. Pharm.*, 528:133–143. DOI: 10.1016/j.ijpharm.2017.05.071. 87

[30] Mohammad Hadi, L., Yaghini, E., Stamati, K., Loizidou, M., and MacRobert, A. J. 2018. Therapeutic enhancement of a cytotoxic agent using photochemical internalisation in 3D compressed collagen constructs of ovarian cancer, *Acta Biomaterialia*, 81:80–92. DOI: 10.1016/j.actbio.2018.09.041. 87

[31] Mohammad Hadi, L., Yaghini, E., MacRobert, A. J., and Loizidou, M. 2020. Synergy between photodynamic therapy and dactinomycin chemotherapy in 2D and 3D ovarian cancer cell cultures, *Int. J. Mol. Sci.*, 21:E3203. DOI: 10.3390/ijms21093203. 87

[32] Martinez de Pinillos Bayona, A., Woodhams, J. H., Pye, H., Hamoudi, R. A., Moore, C. M., and MacRobert, A. J. 2017. Efficacy of photochemical internalisation using disul-fonated chlorin and porphyrin photosensitisers: an in vitro study in 2D and 3D prostate cancer models, *Cancer Lett.*, 393:68–75. DOI: 10.1016/j.canlet.2017.02.018. 88

Authors' Biographies

LAYLA MOHAMMAD-HADI

Layla Mohammad-Hadi is a doctoral graduate in nanomedicine and cancer therapy from University College London (UCL) in the Dept. of Nanotechnology, Division of Surgery and Interventional Science. She graduated in Pharmacology and then completed a Master's degree in Reproductive Medicine and Women's Health. Her doctoral research was conducted in the Dept. of Nanotechnology, Division of Surgery and Interventional Science at UCL and mainly focused on Photodynamic Therapy (PDT) and the use of Photochemical internalisation (PCI) for the delivery of anti-cancer drugs to their target sites of action in various 3D models of breast and ovarian cancer. After the completion of her Ph.D., Layla continued to carry out research on colorectal and pancreatic cancer therapy using nanomedicine and PDT at UCL.

MARYM MOHAMMAD-HADI

Marym Mohammad-Hadi is a third-year Ph.D. student researching in nanomedicine and cancer therapy at University College London (UCL) in the Department of Surgical Biotechnology, Division of Surgery and Interventional Science. The main focus of her doctoral research is the development of a multistimulus-responsive nanoparticulate formulation for sonodynamic therapy and sonochemical internalization in pancreatic cancer. Marym obtained a degree in Pharmacology as well as a Master's degree in Cancer Therapeutics before commencing her Ph.D.

Printed in the United States
by Baker & Taylor Publisher Services